Watching Earth from Space
How Surveillance Helps Us – And Harms Us

Pat Norris

Watching Earth from Space

How Surveillance Helps Us – And Harms Us

 Springer

Published in association with
Praxis Publishing
Chichester, UK

Mr Pat Norris, FRIN, CITP
Byfleet
Surrey
UK

SPRINGER–PRAXIS BOOKS IN SPACE EXPLORATION
SUBJECT *ADVISORY EDITOR*: John Mason, M.B.E., B.Sc., M.Sc., Ph.D.

ISBN 978-1-4419-6937-8   e-ISBN 978-1-4419-6938-5
DOI 10.1007/978-1-4419-6938-5

Springer Dordrecht Heidelberg London New York

Library of Congress Control Number: 2010932126

Springer Science + Business Media, LLC, © 2010

Cover design: Jim Wilkie
Project copy editor: Christine Cressy
Typesetting: BookEns, Royston, Herts., UK

Printed on acid-free paper

Springer is a part of Springer Science + Business Media (www.springer.com)

Contents

Author's preface

Surveillance cameras are difficult to avoid, from the CCTV cameras in streets, lobbies and shops to the cameras at traffic signals and roadside speed traps. There are cameras in outer space watching us too – not as obvious as the street level variety, but there nonetheless – and their number is increasing rapidly as costs fall and more countries want one that carries their flag. Occasionally they are discussed by the media as when Vice President Cheney's home was deliberately blurred in the satellite image on Google Earth. This book catalogs the main reasons why satellites in outer space are watching the earth. Many of the reasons are benign while others are deliberately intrusive. The theme of the book is that by and large Earth-watching satellites are a good thing.

I was overwhelmed by the assistance I received from friends and colleagues, old and new, in collecting the material presented in the book. If it contains errors or ambiguities, it is because of my failure to correctly interpret their material. The long list of those who helped includes (in alphabetical order) Paul Brooks, Simon Casey, Jeremy Close, John Davey, Mark Drinkwater, Roy Gibson, Anders Hansson, Ray Harris, Dave Hodgson, David Keighley, Adam Keith, Bob Kelley, Hans Kristensen, Robert Meisner, Amy Norris, Allison Puccioni, Ray Purdy, Nick Shave, David Southwood, Sir Martin Sweeting, Nick Veck, Joanne Wheeler. Bob Kelley's contribution was particularly important as without him, chapter 7 on the monitoring of nuclear materials, would not have been possible. To them and all the others who helped me without hesitation, I extend my sincere thanks.

I have tried to attribute copyrights for the images used where they were evident. If anyone wishes to claim an image, I will happily amend the appropriate caption in the next edition of the book.

Pat Norris
August 2010

This book is dedicated to Valerie

Figures

Tables

Abbreviations and acronyms

ABAAC	Argentine–Brazilian Agency for Accounting and Control of Nuclear Materials
ACRIM	Active Cavity Radiometer Irradiance Monitor
ADEOS	ADvanced Earth Observing Satellite
AIS	Automatic Identification System
AIT	Aerial Imaging Team
ALOS	Advanced Land Observation Satellite
ARTEMIS	Advanced Relay and TEchnology MISsion
ASTER	Advanced Spaceborne Thermal Emission and Reflection Radiometer
AVHRR	Advanced Very High-Resolution Radiometer
AVNIR	Advanced Visible and Near Infrared Radiometer type
BATMAV	Battlefield Air Targeting Micro Air Vehicle
BIRA-IASB	Belgisch Instituut voor Ruimte-Aëronomie – Institut d'Aéronomie Spatiale de Belgique
CAP	Common Agricultural Policy
CBERS	China–Brazil Earth Resources Satellite
CCD	Charged Couple Device
CCTV	Closed Circuit Television
CGMS	Coordination Group for Meteorological Satellites
CIA	Central Intelligence Agency
CLS	Collect Localization Satellites
CNES	Centre National d'Etudes Spatiales
CNSC	Canadian Nuclear Safety Commission
COMINT	Communications intelligence
COMS	Communication, Ocean and Meteorological Satellite
CoReH$_2$O	Cold Regions Hydrology High-resolution Observatory
CSIST	Chung-Shan Institute of Science and Technology
DC	District of Colombia
DEM	Digital Elevation Model
DMC	Disaster Monitoring Constellation
DMSP	Defense Meteorological Satellites Program

DoD	Department of Defense
DRC	Democratic Republic of the Congo
DRTS	Data Relay Test Satellite
DSP	Defense Support Program
EGNOS	European Geostationary Navigation Overlay System
ELINT	Electronic Intelligence
Elisa	ELectronic Intelligence SAtellite
ELT	Extremely Large Telescope
EMP	Electromagnetic pulse
EORSAT	ELINT Ocean Reconnaissance Satellites
ERTS	Earth Resources Technology Satellite
ESA	European Space Agency
fAPAR	fraction of Absorbed Photosynthetically Active Radiation
FBI	Federal Bureau of Investigation
GAGAN	GPS Aided Geostationary Augmented Navigation
GB	Gigabyte
GCHQ	General Communications Headquarters
GDACS	Global Disaster Alert and Coordination System
GEO	Geostationary Earth Orbit
GMT	Greenwich Mean Time
GOES	Geostationary Operational Environmental Satellite
GPS	Global Positioning System
GRACE	Gravity Recovery and Climate Experiment
GRU	Glavnoye Razvedyvatel'noye Upravleniye (Soviet military intelligence agency)
GSFC	Goddard Spaceflight Center
HIRU	Hemispherical inertial reference unit
HQ	Headquarters
IAEA	International Atomic Energy Agency
ICBM	Intercontinental Ballistic Missile
IED	Improvised Explosive Device
IGS	Information Gathering Satellite
IMS	Indian Mini-Satellite
INPE	Instituto Nacional de Pesquisas Espaciais
INVO	Iraq Nuclear Verification Office
IPCC	Inter-governmental Panel on Climate Change
IRS	Indian Remote Sensing Satellite
ISRO	Indian Space Research Organisation
JAXA	Japan Aerospace Exploration Agency
JPSD	Joint Precision Strike Demonstration
JWST	James Webb Space Telescope
KARI	Korean Aerospace Research Institute
KGB	Komitet gosudarstvennoy bezopasnosti (Soviet security agency)
KH	Keyhole
LDCM	Landsat Data Continuity Mission

LEO	Low Earth Orbit
Lidar	Light detection and ranging
LIMES	Land and Sea Monitoring for Environment and Security
LUSI	Lumpur (mud) Sidoarjo
MIRV	Multiple Independently-targetable Re-entry Vehicle
MODIS	Moderate Resolution Imaging Spectroradiometer
MSAS	Multi-functional Satellite Augmentation System
MSG	Meteosat Second Generation
MTCR	Missile Technology Control Regime
MTSAT	Multi-mission Transport Satellite
MUSIS	Multinational Space-based Imaging System
NASA	National Aeronautics and Space Administration
NASDA	National Space Development Agency of Japan
NASIC	National Air and Space Intelligence Center
NATO	North Atlantic Treaty Organisation
NCDC	National Climatic Data Center
NESDIS	National Environmental Satellite, Data, and Information Service
NGA	National Geospatial-Intelligence Agency
NGDC	National Geophysical Data Center
NGEO	Next Generation Electro-Optical
NGO	Non-Governmental Organization
NOAA	National Oceanic & Atmospheric Administration
NOSS	Naval Ocean Surveillance Satellite
NPIC	National Photographic Interpretation Center
NPOESS	National Polar Orbiting Environmental Satellite System
NPT	Nuclear Non-Proliferation Treaty
NRO	National Reconnaissance Office
NSA	National Security Agency
NSG	Nuclear Suppliers Group
OCHA	Office for the Coordination of Humanitarian Affairs
OECD	Organisation for Economic Cooperation and Development
OPCW	Organization for the Prevention of Chemical Weapons
PALSAR	Phased Array type L-band Synthetic Aperture Radar
PRISM	Panchromatic Remote-sensing Instrument for Stereo Mapping
R&D	Research and Development
REDD+	Reducing Emissions from Deforestation and forest Degradation in developing countries
RORSAT	Radar Ocean Reconnaissance Satellites
SALT	Strategic Arms Limitation Treaty
SAM	Surface-to-Air Missile
SAR	Synthetic Aperture Radar or Search And Rescue
satnav	Satellite navigation
SBIRS	Space-Based Infra-Red System
SDS	Space Data Systems
SIAU	Satellite Imagery Analysis Unit

SSOT	Sistema Satelital para Observaciœn de la Tierra
SSTL	Surrey Satellite Technology Limited
TAS	Thales Alenia Space
TDRSS	Tracking and Data Relay Satellite System
TES	Technology Experiment Satellite
TSF	Télécoms Sans Frontières
TV	Television
UCAR	University Corporation for Atmospheric Research
UHF	Ultra High Frequency
UK	United Kingdom
UN	United Nations
UNHCR	United Nations High Commissioner for Refugees
UNICEF	United Nations Children's Fund
UNMOVIC	United Nations Monitoring, Verification and Inspection Commission
UNOSAT	UN Institute for Training and Research (UNITAR) Operational Satellite Applications Programme
UNSC	UN Security Council
UNSCOM	UN Special Commission
URENCO	URanium ENrichment COmpany
USA	United States of America
USAF	US Air Force
VLT	Very Large Telescope
VoIP	Voice over Internet Protocol
VP	Vice-President
VT	Victoria Terminal (Mumbai)
WAAS	Wide Area Augmentation System
WHO	World Health Organization
WiFi	A play on the term "high fidelity" involving wireless broadband
WMD	Weapons of Mass Destruction
WMO	World Meteorological Organisation

1

The threat of satellite images

SATELLITE IMAGES HIT THE HEADLINES

"Google Earth helps extremists terrorise Mumbai," scream the headlines as more than 170 people are killed in November 2008. The gunmen came ashore on the evening of November 26th and made their way to targets across the south of the city of 13 million.

The attackers all came from Pakistan and were unfamiliar with India's most populous city. One of the gunmen was captured alive, and police say he confirmed that Google Earth was used to familiarize the gang with the streets of Mumbai. Mumbai lawyer Amit Karkhanis filed a petition at the Mumbai High Court calling for Google Earth images of sensitive areas to be blurred so as not to aid future terror groups.[1]

Confirmation of the perceived threat posed by satellite images comes from reports that then US Vice-President Dick Cheney persuaded Google Earth to blur images of his home in Washington, DC.

British soldiers fighting insurgents in southern Iraq are considering suing Google Earth based on evidence that the imagery was being used to target mortar attacks on the army's base near Basra. "The terrorists know exactly where we eat, sleep and go to the toilet," one soldier said.[2]

The addition of the Street View feature to Google Earth has incensed privacy campaigners even more. Satellite images and maps have now been joined by high-quality ground-level photos in many of the areas covered by Google Earth. Greece has banned Google from expanding the Street View service. In Japan, Google has had to reshoot its photos closer to the ground to avoid looking over fences. A formal complaint against Street View in Britain was dismissed by the UK's Information Commissioner, although he ruled that it carried a risk of invading privacy.[3]

---

[1] Blakely (2008).
[2] Harding (2007).
[3] BBC (2009).

P. Norris, *Watching Earth from Space,* Springer Praxis Books,
DOI 10.1007/978-1-4419-6938-5_1, © Springer Science+Business Media, LLC 2010

The Obama Administration has officially confirmed that satellite images are a threat to privacy. The Bush Administration had authorized domestic security agencies to receive spy satellite imagery previously seen only by the military and the CIA. However, Obama's Homeland Security chief, Janet Napolitano, killed that plan in response to concerns from privacy groups. Californian Democratic Congresswoman Jane Harman said the previous Administration's plan "was an ill-conceived vestige of the 'dark side' counterterrorism policies of the Bush years" and "just an invitation to huge mischief". Welcoming the decision to halt the domestic use of spy satellites, she said "it showed real leadership on the part of [Homeland Security Director] Janet Napolitano".[4]

It seems that satellites can spy on us from space and pose a threat to our safety and privacy.

But how real is this threat? Could it be media exaggeration or political spin?

Let's examine each of the stories above – Mumbai, Dick Cheney's house, the Iraq mortar threat and Janet Napolitano's action – to see how much is hype and how much is reality.

First, let's understand what satellites can and can't see from space.

WHAT CAN A SATELLITE SEE FROM SPACE?

Satellites fly across the sky unimpeded by borders, typically 200–800 km above the ground. With a camera onboard, they can take photographs of the ground below (if there are no clouds in the way). Many satellites carry cameras for benign reasons, such as weather forecasting or environmental monitoring, but some are explicitly seeking military information, and these we call spy satellites. If they have a suitable radio receiver, they can listen in to whatever radio signals are being transmitted below and this sort of satellite usually has a military objective. In this part of the story, the camera-carrying satellites are the main focus of the story but I will return to the radio listening satellites in Chapter 9.

Modern spy satellites are like a camera-equipped cell phone – with a very long telescope attached. You take a photo of an interesting scene using the built-in camera of your cell phone. The image is stored in the computer-style memory of the phone. Then, you send it to whoever is interested in it via the cell phone network – in other words, by radio link.

The number of pixels in the image taken by the phone-camera dictates its graininess or resolution, namely the ability to blow it up and see further detail. The more pixels in the camera, the more you can magnify it without it becoming grainy. On the other hand, the more pixels in an image, the fewer the images you can store in the memory and the more you pay to transmit it over the network. You can zoom in on the subject before taking the photo, in which case you will see more detail in the final image but less of the surroundings.

---

[4]   Sullivan (2009), and Hsu (2009).

Sometimes, there is no network coverage where you have taken the picture and you have to wait until your journey brings you to an area with network coverage before you can send it. If this is a frequent occurrence, you may consider switching to another network that offers better coverage.

This same outline could as easily apply to a modern spy satellite, leaving aside the fact that it takes its pictures at a distance of 200 km or more through a telescope. It takes images – automatically to a predetermined schedule rather than when a human touches a button. The heart of the camera is the same technology as in the digital camera or cell phone you buy in the high street – probably a charged couple device (CCD), which is a form of solid-state electronics similar to the transistor and the computer chip, which turns light into electrical messages. The images are stored in computer memory on the satellite, which can only hold a certain number of images before it reaches its capacity. The images may be radioed to ground immediately, but frequently there isn't a friendly ground station within sight of the satellite so it waits until its orbit brings it within the coverage of its ground network.

The USA and Russia have installed extra network relay stations to improve the coverage and thus get images back to ground from more or less anywhere in the world. The relay stations are actually satellites located at suitable very high orbits that relay the images from the spy satellite to the relevant ground network. These satellites therefore have no intrinsic limitation as to the number of images they take, provided the radio link has enough capacity to carry them. Other spy satellite operators such as France, China and India have to wait until their satellites appear over their ground stations before receiving the recorded images.

The atmosphere limits what spy satellites can see in several ways. Cloud cover prevented the early spy satellites from taking useful photographs much of the time. It would be the late 1980s before the first radar spy satellite was in orbit able to see through the clouds – more about this later (Chapter 8). Even when the sky is clear of clouds, the hours of darkness prevent detailed photographs being taken. Most of the Soviet Union is north of 50° latitude and much of it above 60°. The good news for American spy satellites was that the summer provided long hours of daylight, but the bad news was that winter nights were equally long. Paranoid American military analysts worried about what the Soviets might be up to during those long hours of darkness. By contrast, all of the continental USA is south of the 49th parallel, making the summer/winter contrast less extreme for Soviet spies in the sky.

Even when clear and in daytime, the atmosphere is turbulent. We see this for ourselves when viewing the stars – the twinkling of starlight is due to the shimmering of the atmosphere. This atmospheric turbulence puts limits on the accuracy of spy satellites. Although accuracy figures for the latest spy satellites are secret, we can work out the accuracy achievable with the Hubble Space Telescope if it were to point at the earth, which will give us a rough idea of what is possible. Hubble was designed to have a resolution of about 15–30 millionths of a degree.[5] If the atmosphere were

---

[5]   Chaisson (1998) p. 29.

completely still and Hubble was pointing at the earth from a typical spy satellite altitude of 250 km, it could resolve objects on the earth's surface of about 5–10 cm (2–4 inches) in size.

What do we mean by a resolution of 10 cm? It doesn't mean that two objects 10 cm apart will always be recognized as two objects and that two objects 8 cm apart will always be recognized as a single larger object. The ability to detect the small gap between two objects will depend on factors such as lighting conditions, the shapes and surfaces of the objects, shadowing in the gap, and the color and sheen contrast between the two objects and between them and the gap. It means that, in general, objects 10 cm apart will be recognized as being separate, while objects 8 cm apart will be more likely to be recognized as a single object. It also means that a gap of 1 cm, say, between two objects would hardly ever be detected.

Telescopes can focus an image onto a film down to a limit set by the color (wavelength) of the light. The resolution of the spy satellites comes down to how sharply you can focus light onto the film and the inherent graininess of the film itself. Early spy satellites generally worked at the limit of both of those parameters – the smallest image picked up by the telescope roughly equal to the graininess of the film.

Digital cameras and the camera in our cell or mobile phone have familiarized us with the term *pixel* or picture element. A camera that takes images containing 5 million pixels gives better pictures than one that contains 2 million pixels. If you blow up the picture, the graininess in the 2-million-pixel image becomes evident, whereas the 5-million-pixel image still looks sharp. When graininess appears, you have reached the ultimate resolution of the picture – objects smaller than the graininess are blurred and can't be resolved.

Even if a surveillance satellite has a resolution of, say, 10 m, it may be possible to detect and interpret features smaller than that. A satellite image with 10-m resolution of a 15-m-wide road may well show the white lines along the center of the road. The white line is only 5 cm wide so you could argue that the image has 5-cm resolution. This illustrates that the contrast in brightness between two objects (the 15-m-wide road and the 5-cm-wide white line) can make an object visible even though it is much smaller than the theoretical resolution of the image. The figure of 10-m resolution in this case is presumably based on some sort of average conditions.

An example of getting information that is better than the resolution in the military sphere is to work out the width of a missile to see whether it complies with a Treaty agreement. It should be possible to tell this with an accuracy as much as 10 times better than the image resolution, because the rim of the missile is made up of several pixels, giving us an averaging effect. I can speak for this personally through two commercial systems that I have been involved with in recent years.[6] In the first one for a Japanese customer, our software compares features in a satellite picture of the earth with features stored in a computerized digital map. The purpose is to work out from this comparison how much the image is distorted or the satellite is mis-

---

[6]   At my employer, Logica plc.

pointing. The resulting accuracy is 10 times better than the size of an individual pixel. The second system was for a European customer and achieved similar sub-pixel accuracy in monitoring the movement of clouds from one satellite image to the next – thereby measuring the speed and direction of the wind (more about this particular topic in the next chapter).

So, the quality of a spy satellite image is not a simple resolution value in meters or centimeters. A sophisticated scale of quality from 0 (worst) to 9 (best) was defined by the National Reconnaissance Office in terms of the information you could obtain from the image. For example, to say that an image was level 4 meant that you could see whether the door of a missile silo was open or closed and was equivalent to a resolution of about 2 m. At level 6 or about 50-cm resolution, you could distinguish between several different types of missile. We will return to this topic in Chapter 8, in which Figure 109 outlines the military view of image resolution.

During the Cold War, Robert Kohler, then at the CIA, recalls that in addition to defining what each quality level was in words, they tried to have an image that illustrated each level. It proved difficult to find an image, even a low-level airborne image, with level 9 quality – the highest possible quality level. Finally, a picture taken by an aircraft flying along the border between East and West Germany proved to have the required quality – it showed an East German soldier urinating. The image was displayed for all to see under the banner headline "German soldier pissing in the snow – level 9".[7]

The four pictures in Figures 1, 2, 3 and 4 are courtesy of John Pike and his GlobalSecurity organization.[8] They show two newspaper front pages, a vehicle license plate and a golf ball in images of various resolutions. Figure 1, with a resolution of 10 cm, is probably the best that spy satellites can do and doesn't come close to reading the headlines or the number plate nor even detect the golf ball – no chance to use these satellites to find that golf ball in the rough, then! Figure 2 shows the quality available from some aerial imaging systems and this probably can detect the golf ball but nothing else. Figure 3, with a resolution of 1 cm, allows you to read large tabloid headlines but not the license plate or normal headlines – but, interestingly, does resolve the picture on the *New York Times* front page quite well. Finally, Figure 4, with 1-mm resolution, which is way beyond the capability of current satellites, shows what the other images miss – the golf ball is distinguished as circular and not a small white box, the license plate is revealed and headlines, pictures and text in both newspapers are legible.

The Hubble Space Telescope was developed more than 20 years ago, so we might expect current spy satellites to be somewhat better in their performance. But the fact remains that resolving from space details on the surface of the earth that are smaller than a few centimeters is almost impossible. Forget about recognizing Osama Bin Laden as he walks down the street (even if he conveniently looks up to the sky at the

---

[7]   McDonald (2002) p. 223.
[8]   *www.globalsecurity.org.*

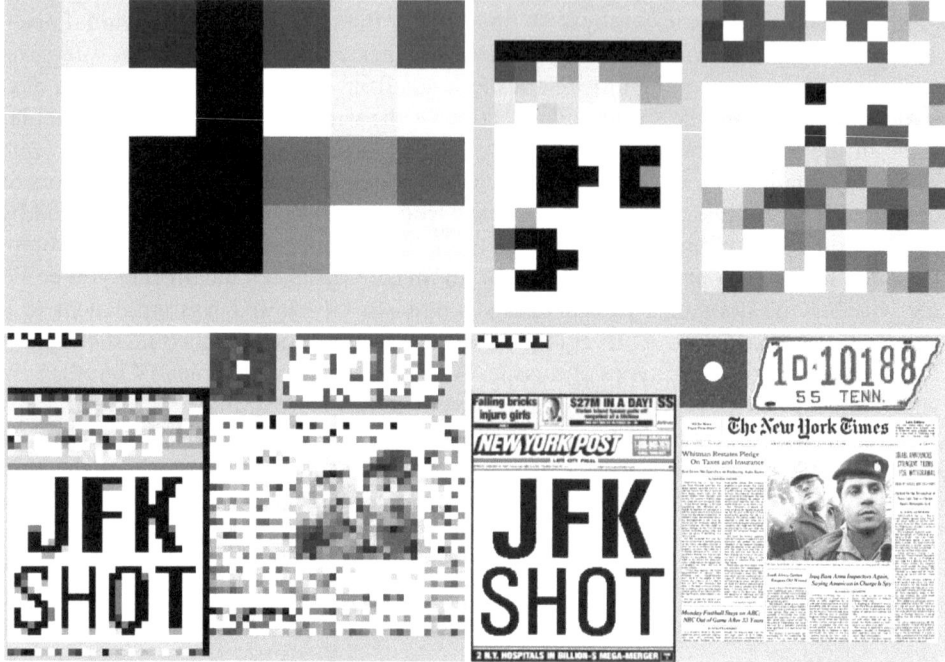

Figures 1, 2, 3 and 4. Simulation of images with various resolutions: 10 cm (Figure 1, high-quality spy-satellite image), 3 cm (Figure 2, high-quality aerial photo), 1 cm (Figure 3) and 1 mm (Figure 4). Courtesy John Pike and *GlobalSecurity.org*.

right moment!) or reading car license plates – or the newspaper headline. Photos that have been released show that it might occasionally be possible to identify the make and model of a car.

And all of that assumes that the atmosphere is absolutely still – which it hardly ever is. The atmosphere is constantly moving due to thermal gradients within it. On a really hot day, the shimmering is visible to the naked eye, for example, above a tarmac surface or a hot sandy beach. As you look further and further through the air, the thermal shimmering accumulates, so that through a telescope or strong binoculars, the shimmer is quite apparent. Even in the cool of the night, the shimmering air causes the stars to twinkle. In astronomical photos, the atmospheric shimmer is about 1 second of arc, which, in a photo of an object 100 km distant, is about 50 cm. Thus, because of this shimmer, an astronomical telescope photographing a satellite overhead could not make out features smaller than 50 cm – and more or less the same applies in the reverse direction with a spy satellite taking photos of the ground. The technology in a modern digital camera that removes jitter or focuses on eyes can get around this air shimmer to some extent. The software in a digital camera detects each jitter event by seeing objects in the picture move and moves the camera lens in the opposite direction. Satellite cameras have always done something like this – in the one-thousandth of a second (a millisecond) needed to take an image, the satellite has moved across the ground by about 8 m.

Satellite designers compensate by moving a mirror inside the camera in the opposite direction to avoid what would otherwise be a blurred image. Overcoming air shimmer takes this idea a bit further in that software in the satellite would find a suitable object in the picture – the edge of a building or a river or any sharp feature – and if it moves more or less than expected, adjust the mirror movement accordingly. Astronomers on earth are using this concept to take images as sharp as those of Hubble in space, so it seems likely that military imaging satellites are doing the same from space.

Legislators have tended to focus on resolution when formulating laws on satellite imaging. They have generally ignored other features of the satellites that can be just as important as resolution in allowing objects to be detected and analyzed. These other characteristics include the ability to distinguish color – or spectrum, to give it a technical name. Color in the images (as opposed to black and white) makes it possible to distinguish between crops, for example, and how healthy they are. It may also allow you to analyze the plume emitted by a rocket and thus tell what fuel it is using. The ability to analyze the brightness of an object is another important feature. An image that distinguishes subtle shades of gray will tell us much more than a simple black-and-white picture. The resolution of an image can be improved by taking two or more photos of the scene from slightly different angles. Combining the images will provide an image that has better resolution than one image on its own – provided the scene hasn't changed in the meantime. A single satellite with a simple camera may take an image of the same scene for several days; thus, if a satellite or group of satellites has the ability to take a repeat image within a short time (seconds or minutes or even hours), that is important. A variation on this revisit feature is to combine two images to give a stereo view of the scene – features not evident in either image can stand out, literally as well as metaphorically, in the stereo image. These other capabilities are not covered in current legislation of surveillance by satellite, thereby weakening what legislation there is on the subject.[9]

THE MAKINGS OF A SATELLITE

Let's take a look at a typical imaging satellite.

The artist's impression of the 4-ton JAXA (Japan's space agency) Daichi satellite launched in January 2006 (and previously called ALOS) in Figure 5 shows many features typical of all earth-observing satellites. The satellite is dominated by the elongated solar array panel on the right. The sun-facing side of this panel is covered with solar cells that convert sunlight into electricity. Daichi's solar array is 22.2 m (73 ft) long and 3.1 m (10 ft) high. It is made up of nine panels that are folded concertina-fashion for the launch (on the left in Figure 6) then unfold once the satellite is in orbit. You may have seen pictures of the International Space Station,

---

[9]   Hanley (2000).

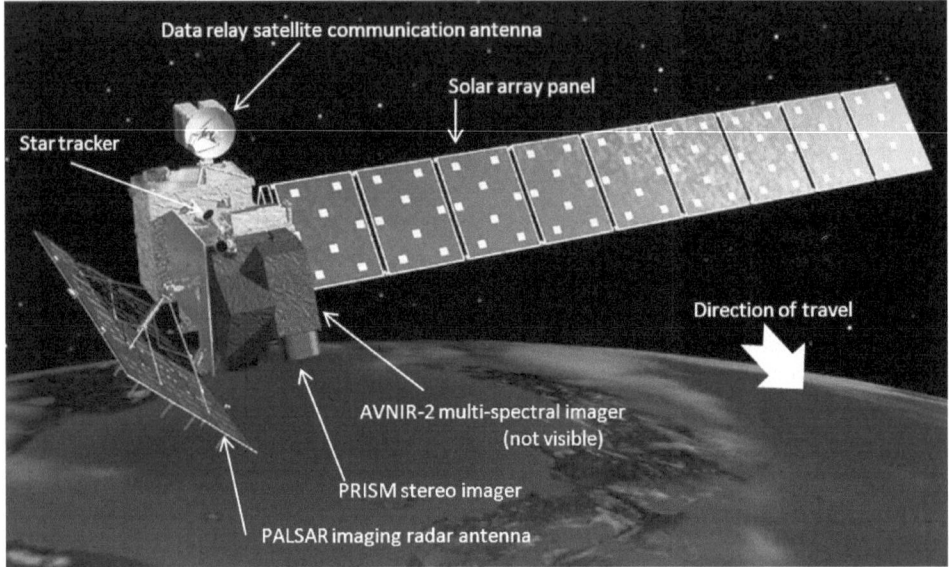

Figure 5. Artist's impression of Japan's Daichi satellite showing the main features of a surveillance satellite. Courtesy JAXA.

which is also dominated by its several solar arrays, indicating the importance to all satellites of a sustainable source of electrical power – this allows them to stay in orbit for many years.

On the left of the artist's impression (Figure 5) is another panel, which is the antenna for the imaging radar – the radar is sufficiently complex and sophisticated to be given a name of its own: PALSAR.[10] It is almost 9 m (29 ft) long and 3.1 m (10 ft) high. Although the eventual information from PALSAR is images, it is not a camera or telescope. It forms images by measuring the individual echoes of its radar transmission from each point on the ground below. Each individual echo is then considered as a pixel to form an image. The transmission and returning echo are radio signals, hence the need for an antenna.

Daichi carries two other imaging instruments. The PRISM[11] instrument is a triple camera that can take stereo images using three telescopes that point towards the ground at slightly different angles – you can just make out two of the three telescopes in the image. One points straight down and the others point forward and backward. The three images are combined on the ground to create stereo imagery, which is helpful for analyzing ground contours. The resolution of the images is 2½ m directly below the satellite but tails off as you look to either side.

---

[10]   PALSAR is an acronym of Phased Array type L-band Synthetic Aperture Radar.
[11]   PRISM = Panchromatic Remote-sensing Instrument for Stereo Mapping.

Figure 6. Daichi before being launched. Courtesy JAXA.

The final imager is called AVNIR-2,[12] which takes images in four colors, one of which is in the infrared and detects heat – this means that it can take images at night, with cold objects appearing black and warm ones bright. To take the color images, AVNIR-2 is effectively made up of four separate cameras, one for each color – the cameras all stare through the same telescope thanks to a clever arrangement of mirrors that focuses the scene onto all four cameras simultaneously. The images have a resolution of 10 m.

The main body of the satellite in Figure 5 is roughly a rectangular box, 6.2 × 4 × 3.5 m. It is packed full of fuel tanks, electronics, batteries and the like. The fuel feeds its small rocket motors that keep it pointing in the right direction and make small changes to its orbit from time to time. The batteries are charged up when the satellite is in sunlight and are used to power the equipment on the satellite during the periods (sometimes a third of the orbit) when the sun is hidden behind the earth. The star tracker shown on the top of the main body ensures that the satellite is pointing in the right direction – it takes images of the dark sky overhead (the sky is always black in outer space) and identifies the brightest stars in its field of view so that it knows which way is up.

Above the main body is a small dish antenna that communicates with a satellite in geostationary orbit (36,000 km above the earth). By having this dish, Daichi is able to immediately send its images to the ground when it is in view of the relay satellite. Without it, Daichi would have to store the images in its onboard memory and wait until a suitably equipped ground facility was in view. The geostationary satellite normally used by Daichi is Japan's Data Relay Test Satellite (DRTS, also called Kodoma), which is specially equipped to receive the transmission from satellites like Daichi. Kodoma is an experimental satellite and a replacement is not currently funded. Located over the Indian Ocean, it can pick up data when Daichi is over Asia, the Indian Ocean and the eastern Pacific Ocean. To deal with imagery of other parts of the globe or if Kodoma is not available, Daichi also has the ability to work in the conventional way – storing onboard and transmitting when a station is close.

In 2010, Daichi started sending images to earth through NASA's TDRSS relay satellites (see below), thus doubling the number of images of the western hemisphere that it can provide.[13] Europe's ARTEMIS[14] is similar to Kodoma and is used by other surveillance satellites such as France's SPOT satellites.

Not many satellites have the ability to communicate with a geostationary satellite like this. In the past, the USA and Russia were the only countries with the technology to do it. The USA has a fleet of eight Tracking and Data Relay Satellite System (TDRSS) geostationary satellites for this purpose: TDRSS-C through -J (TDRSS-A was switched off in June 2010 after 27 years in use; TDRSS-B was destroyed in the explosion of the Space Shuttle Challenger in 1986) (see Figure 7).

---

[12]   AVNIR-2 = Advanced Visible and Near Infrared Radiometer type 2. AVNIR-1 was onboard the ADEOS satellite launched in 1996.

[13]   Morring (2010a).

[14]   ARTEMIS = Advanced Relay and TEchnology MISsion.

Figure 7. Artist's impression of four of NASA's TDRSS satellites in orbit. Each of its big dishes can communicate with a separate satellite in orbit far below. Credit: NASA.

The TDRSS satellites act as relays for transmissions from NASA vehicles and satellites such as the Space Shuttle, the International Space Station, the Hubble Space Telescope and various earth-observing and scientific satellites. The two big dishes on TDRSS are 4.9 m (16 ft) wide and each can lock on to a separate satellite – Japan's Kodoma and Europe's ARTEMIS can only cope with one satellite at a time.

The advantage of the TDRSS group of satellites is that they see a satellite in a low orbit for more than 85% of that orbit whereas each ground station would see the low-orbiting satellite for only about 10% of the orbit – so TDRSS is as good as eight or nine stations and avoids the need to find willing countries suitably located to give continuous coverage of the orbiting satellites. The disadvantage of TDRSS is that each TDRSS is pretty expensive and each satellite that uses it has to carry an antenna that can lock on to TDRSS and steer itself to stay locked on as it moves in its orbit. Furthermore, TDRSS itself needs a ground station, and a particularly complex one at that, since it is effectively dealing with dozens of different satellites around the world. The complexity of the station was brought home to NASA when an upgrade in the 1990s cost a lot more and took a lot more time than was intended.

Unlike NASA, the commercial world has tended to prefer the ground station option rather than the geostationary relay option. Commercial satellites can usually afford to wait a few hours to get their images back, whereas NASA wants instant and continuous links to the Space Shuttle and the International Space Station. Commercial satellites therefore benefit from the recent advances in mass storage –

consider the number of images you can now store on a tiny chip in your digital camera compared with just a few years ago. The satellites store the images until they are in sight of the station. The commercial operators also benefit from the trend to leave ground stations virtually unmanned – with maintenance staff visiting perhaps once a week and security provided by basing the station in a larger complex, such as a commercial park or a military base.

Mind you, NASA's Landsat satellites have been thankful that they were equipped to send their images back through TDRSS. The two radio transmitters on Landsat 4 for sending images direct to ground both failed, leaving TDRSS as the only way to get its images back.

The US military has its own relay satellites called SDS-2 and QUASAR[15] in high orbits, not all geostationary, allowing them to relay images from surveillance satellites over the poles, which would be out of view of a geostationary satellite. Some US military satellites also use TDRSS. Russia has similar relay satellites for its military surveillance satellites. This topic is discussed further in Chapter 8. Although most of the remainder if this book is about why satellites are watching the earth, I will return to how satellites work from time to time when a specific feature not described above is being discussed. You can find these sections in the index at the end of the book under "How satellites work".

Let's now turn to the topics mentioned at the start of the book, beginning with the terrorist attack in Mumbai.

MUMBAI ATTACKS, NOVEMBER 2008

The 10 Pakistanis who came ashore on the evening of November 26th 2008 may never have been to Mumbai, India, before, but their trip had been prepared by others who had. One hundred and seventy-five people died in the 3 days of terror, including all but one of the terrorists. The terrorists took cabs from where they came ashore to the five target locations, thereby not requiring any great geographical knowledge. A detailed dossier of their movements and planning was compiled by the Indian Government based on information from the one captured terrorist and physical evidence such as the logs in the four GPS satellite navigation devices recovered (one on their boat), intercepted phone calls, and phones and documents found on the terrorists and in their boat.[16] The dossier was given to the Pakistani Government to persuade them to arrest several Pakistani residents identified by India as being implicated in the plot.

The group left Karachi in Pakistan on November 22nd 2008, 550 km from their destination, and hove to 7 km off the coast of Mumbai about 4 p.m. on Wednesday November 26th. They waited until darkness, killed the captain of the boat (which

---

[15]   SDS-2 = Space Data Systems second generation.
[16]   Indian Government (2009).

they had hijacked near Pakistan) and came ashore in an inflatable dinghy at about 8:30 p.m. on the western side of Coloba Causeway close to the commercial and tourist heart of the city. They then split into five groups of two, each group taking a cab to the five target locations: the Taj Mahal Hotel, the Oberoi Trident Hotel, the Leopold Café, the Nariman or Chabad House Jewish outreach center and the Chatrapati Shivaji Terminus (commonly called "VT", based on its original name, Victoria Terminus). The groups appear to have staggered their departure from the landing area, since their arrival times at their targets ranged from 9:20 p.m. to 10:20 p.m. – the first attack taking place at VT, the farthest from their landing point.

Except for Nariman House, which apparently was targeted because of its Israeli and/or Jewish connections, all of their destinations are well known and well signposted landmarks or tourist spots, and would have been easy to recognize. The rich extravaganza of domes, spires and arches of VT station, for example, led railway historian Christian Wolmar to designate it "the world's most ostentatious and magnificent railway station".[17] The two hotels on the list were known to be frequented by local and foreign politicians and celebrities, as was the 140-year-old Leopold Café.

Each terrorist was armed with a Kalashnikov AK-47 rifle, a hand-gun, about a dozen grenades and a small bomb (described as an Improvised Explosive Device, IED, in the Indian dossier).

The death toll at each location was horrific and the list of seriously injured runs into many hundreds.[18] Fifty-eight people died at VT station in a hail of indiscriminate gunfire and grenade blasts, and seven police officers and one of the terrorists were killed in the chase and confrontation that followed. The second terrorist, Ajmal Kasab,[19] was captured. Ten people were killed at the Leopold Café from a grenade and Kalashnikov bursts – police recovered five empty or near empty AK-47 30-bullet magazines at the scene. The two terrorists then ran one block to the nearby Taj Mahal Hotel, where they joined up with another group. In a 2½-day stand-off involving many hostages, gunfire, grenades and fires lit by the terrorists, 32 guests and staff plus one Army Major in the response team were killed – 20 died in the initial hail of gunfire in the entrance lobby. All four terrorists were killed. The death toll was similar at the Oberoi Trident Hotel, where the terrorists held out for nearly 2 days. Thirty-three guests and staff died at the Oberoi Trident plus both terrorists. At Nariman House, five staff and guests died plus one soldier in the response team and both terrorists. The gunmen fired at local residents and bystanders from Nariman House, leading to at least one tragic incident when an escaping Israeli guest was mistaken by neighbors as a gunman and badly beaten.[20]

---

[17] Wolmar (2009) p. 267 and figure.

[18] I have used the casualty figures in the Indian Government dossier. Figures in the media vary and differ slightly from those in the dossier but do not change the scale of the tragedy.

[19] The dossier gives his full name as Mohammed Ajmal Amir Kasab but media reports shorten it to Ajmal Kasab.

[20] Gee (2009) p. 28.

Bombs left by the terrorists in two of the cabs went off some time later (one near the airport, the other near the port), killing the drivers and passengers in both plus a number of bystanders. Another bomb left by the Nariman House attackers exploded in a nearby petrol station. The random locations of these bombs led the detective directing operations at the police HQ, Joint Commissioner Rakesh Maria,[21] to say "we felt the whole city was under siege, under attack".

Newspaper reports around the time of the attack referred to Google Earth, with its satellite imagery being used by the terrorists to familiarize themselves with the streets of Mumbai. The Indian dossier makes no mention of Google Earth but does implicate two other advanced technologies.

Four standard mass-market satellite-navigation devices were found by police and these would have been more useful than Google Earth in aiding the terrorists. This is the technology that guides hundreds of millions of cars, bicycles and pedestrians to destinations all over the world. Each device contains a digital road map of one or more countries, displays the location of the device using GPS satellites and issues instructions on how to reach the desired destination. Tourist hot spots like Leopold Café, the Taj Mahal Hotel and VT station would be explicitly marked. You don't need to recognize streets or landmarks; you just follow the instructions: "turn right", "turn left", etc. We will look at this technology in more detail in Chapter 6.

The second technology is two modern forms of the telephone: the cell phone, including a relatively unusual version that uses satellites, and the internet phone. The Indian dossier states that the police intercepted calls between the terrorists and various callers. The callers used an internet phone service called VoIP[22] so that the origin of the calls could not be traced. The nearest the dossier gets to identifying the location of the callers is to assert that the activation charges for the internet phone line were paid via Western Union money transfer by a Pakistani national. We will relate how VoIP is the bane of Western security services in Chapter 9. The satellite phone is a device slightly larger than a standard cell phone that communicates through the Thuraya satellite. Thuraya is a commercial service operated from Abu Dhabi in the (Arabian) Gulf available across south, east and west Asia, Africa and Europe. Although the published intercepted calls were all on normal cell phones, cryptic text messages were recovered from the Thuraya phone that had been sent to various other Thuraya phones – the dossier claims that one of these others belongs to a senior official in a Pakistani-based militant Islamic group called Lashkar-e-Taiba (Army of the Righteous), whose leader, Hafiz Saeed, had previously encouraged suicide attacks, stating that "a suicide attack is the best form of Jihad".[23]

The story as it has emerged suggests that Google Earth and satellite images played little or no part in the Mumbai attacks. The terrorists used other advanced technologies, including internet telephony and GPS handsets, to help achieve their objectives.

---

[21] He has since been made chief of the Maharashtra State Anti-Terrorist Squad.
[22] VoIP = Voice over Internet Protocol.
[23] Rashid (2009) p. 228.

VICE PRESIDENT CHENEY'S HOUSE, IRAQ AND THE DOMESTIC USE OF SPY SATELLITES

At the start of this chapter, four high-profile stories were cited as evidence that satellite imagery was a danger to the public. The Mumbai attack was one and we have seen that satellite imagery was irrelevant to the tragic outcome of that event. Let's see whether the other three stories prove different.

The second newspaper headline related to Vice-President Cheney arranging for his residence to appear as a blur in the satellite images on Google Earth. The Vice-President lives in the US Naval Observatory grounds near the up-market Embassy Row in North West Washington, DC. Google denied that it had pixellated or blurred the VP's residence and said that other online mapping services such as Yahoo and Microsoft gave the same level of detail – a view challenged by several online commentators. In any case, in the week that Vice President Biden took office, the view on Google Earth became a lot clearer. A Google spokeswoman said that the change was part of a routine and coincidental upgrade of imagery in the Washington, DC, area. She also said that "some of [our third party suppliers of overhead imagery] may blur images before they provide them to Google".[24] Google says that its latest Washington, DC, imagery is supplied by GeoEye, one of the new breed of commercial high-quality satellite images (discussed further in Chapters 4 and 8). GeoEye Vice President Mark Brender denies that they are required to blur sensitive areas.[25]

Imagery on Google, Microsoft or Yahoo is not current and can be months or years old. The suppliers of satellite imagery generally have a policy of delaying the provision of ultra-detailed images for at least a day – this is mandatory for US companies like GeoEye as a provision of their license to operate and voluntary for non-US companies that want to stay friendly with Uncle Sam, if they want to sell their images to the US Army, for example.

A quick browse through the online imagery websites such as Google Earth shows that the White House and the VP's home are slightly blurred compared with housing 100 m away. But the Pentagon, home of the Defense Department, and the CIA headquarters, both just across the Potomac River, are clear as a bell. Even the headquarters of the super-secret National Security Agency a few miles north of the Capital Beltway in Maryland is not blurred. It is hard to see how security reasons would justify blurring the homes of the top politicians but not the country's top military and intelligence centers. So, the slight blurring of the White House and the US Naval Observatory is something of a mystery.

The British Army complaint about Google Earth being used by Iraqi insurgents to target mortar attacks on their Basra base has some similarities with the Cheney story. The online images do indeed show the general layout of buildings and roads

---

[24]   Silva (2009).
[25]   Weinberger (2008).

on the base – but from some time in the past. Hundreds of local Iraqis work on the base, providing support services such as catering, cleaning and transport. Many of them have mobile phones with cameras, so any insurgency group looking for up-to-date information on the base layout has only to seek the help of a few of these local support staff.

It is a fact of modern military life that all activities have to be camouflaged to avoid overhead imagery, if not to fool Google Earth, then to avoid the imaging satellites of China or India or Russia or Algeria, etc. – 30 + countries operate such satellites and the list gets longer each year (more on this in Chapter 8). If the British Army has failed to observe this basic rule in Basra, it can hardly blame Google Earth.

The US Navy was caught out in 2007 when a photo of a nuclear submarine appeared in a naval dockyard scene on Microsoft's Virtual Earth. The photo was taken by an aircraft rather than a satellite and showed the normally hidden propeller that helps keep the submarine silent when it's on patrol. "That propeller is a national secret" was how a military analyst described the image in *Navy News* magazine.[26] Images of the US military build-up for the 2003 Iraq War were readily available on the internet and gave rise to several news reports. Images on the internet are just the tip of the iceberg, since satellites belonging to governments not always friendly to the USA can take similar images at any time and place of their choosing.[27]

The final news headline mentioned at the beginning of the book was that the Obama Administration has cancelled a plan to make military surveillance satellite information available to Homeland Security agencies. During the George W. Bush Administration, Arizona Governor Janet Napolitano had advocated using advanced security technology as a law-enforcement tool. She was presumably comfortable therefore with the signing into law in September 2008 by President Bush of a Bill that extended the domestic use of military satellite imagery for 6 months. The Bill was watered down by Democrats to focus only on emergency response and scientific needs. Homeland-security and law-enforcement surveillance were excluded for the moment.

Now Homeland Security Director in the Obama Administration, Ms Napolitano has been subject to continued Democratic lobbying to ban the domestic use of the military imagery. The critics were concerned that the practice infringed the general principle enshrined in the Fourth Amendment of the US Constitution that guarantees "freedom from illegal search and seizure". That criticism was strengthened by a General Accounting Office report that said the Homeland Security Department "lacks controls to prevent improper use of domestic-intelligence data by other agencies". Taking a different tack, Los Angeles Police Chief William Bratton,[28] representing the Major Cities Chiefs Association, said that

---

[26] Scutro (2007).

[27] Prober (2003).

[28] In the fall of 2009, Bratton retired from the LAPD and became Chairman of New York-based Altegrity Risk International that is bidding to train police forces in Afghanistan and other post-conflict nations – Berfield (2010).

using military satellites for domestic purposes would violate the Posse Comitatus law, which bars the use of the military for law enforcement in the USA.

California Congresswoman Jane Harman, who heads a homeland-security subcommittee on intelligence, said "Having learned my lesson" with the National Security Agency's warrantless-surveillance program, "I don't want to go there again unless and until the legal framework for the entire program is entirely spelled out". Harman was referring to the use of eavesdropping satellites to intercept phone conversations without legal authorization, which had been a common practice during the Bush Administration's first term – and is described further in Chapter 9.[29]

So the politicians weren't criticizing the use of spy satellite imagery within the USA per se, just the lack of practical legal safeguards about its use. For the moment, the Homeland Security Department is concentrating on building relationships with state and local officials – shown to be sorely lacking during the Hurricane Katrina crisis. Congresswoman Harman and the other privacy advocates appear to be unaware that the distinction between civil and military surveillance satellites is becoming blurred (see Chapter 8) and Chief Bratton's remark about violation of the Posse Comitatus Act may yet lead to legal action based on complaints about imagery from satellites such as GeoEye.

The above examples confirm that Google Earth and its Microsoft, Yahoo and other equivalent websites make it easy to locate satellite and aerial imagery of a location of interest anywhere in the world. However, the imagery is of unknown age so you can't be sure that what it shows is still accurate. You can purchase recent satellite imagery on the internet from a number of sources and this is probably more of a threat to security than Google Earth. You can even order a new image to be taken at a certain time or as soon as possible and you will receive it within hours or perhaps a day or so of the image being taken – more expensive than purchasing an existing image but well within the financial reach of many terrorists organizations. So, just as the internet makes it easy to buy books, music and groceries, the same is true of buying satellite images of a quality that 20 years ago would have been highly classified. Donald H. Rumsfeld, Defense Secretary as the USA prepared to invade Iraq, acknowledged the problem but could only say "I wish we didn't have to live with it".[30]

PRIVACY

There's a Latin phrase that enshrines a home-owner's rights to the ground below and the air above: *ad coelum et ad inferos*, that is "upwards to heaven or the sky and downwards to hell or the centre of the earth". If this legal principle is valid, you own the minerals in the ground under your garden or field and you can exclude airplanes

---

[29]   Gorman (2008), and Gorman (2009).
[30]   Umansky (2002).

from flying over your house. If your rights extend all the way to "heaven", as the phrase states, then satellites need permission to overfly your house.

Sadly, the *ad coelum et ad inferos* phrase has little or no validity in law. In most countries, the government restricts your right to drill for oil or mine for minerals perhaps by requiring you to get a license. Likewise, governments allow airplanes to fly across the sky subject to certain regulations about noise and safety. So, it should come as no great surprise that you have little chance of legally stopping a satellite from flying above your property.

However, there are some legal arguments to consider. The UN has agreed a set of principles for surveillance satellites.[31] They are not legally binding in the usual sense, but since no country has legally objected to them, they carry weight at least as a statement of common practice – or of "customary law". The UN principles say that surveillance "shall not be detrimental to the interests" of the country being observed.[32] This means presumably that they shouldn't invade your privacy or threaten your safety. It seems therefore that if you consider another country's satellite is detrimental to your interests, you could ask your government to take action against the owner of the satellite in question. This particular legal avenue appears not to have been tested yet.

The courts in several countries have ruled on the admissibility of satellite images as evidence, with mixed results. The US Supreme Court ruled in 1986 against Dow Chemical in allowing the Environmental Protection Agency to use aerial photos to monitor Dow's facilities. The Court decided that aerial photography was sufficiently mundane that Dow had no reasonable expectation of privacy from it – the ruling stated that aerial photography was not intrusive if details as small as ½ inch (1 ¼ cm) in size could be made out under magnification. However, the Judges explicitly noted that satellite photos might be considered as "highly sophisticated and not generally available to the public" and therefore might require a warrant before being allowable in evidence. Aerial photography was perhaps considered routine in 1986 because some of the Supreme Court Justices were licensed pilots. That was a quarter of a century ago and nowadays you can order satellite photos via the web from a variety of suppliers, which would seem to make them "generally available to the public" and thus admissible as evidence without the need for a warrant.

In 2001, the US Supreme Court declared in a narrow 5:4 decision that thermal imaging of a home from outside was unconstitutional without a warrant because the technology in question is not in general public use. The Court expanded the scope beyond thermal imaging to "sense-enhancing technology". The case involved a Mr Kyllo growing marijuana in his home in Florida. Thermal imaging showed that Kyllo's garage was emitting more heat than was normal but was typical of the high-intensity lamps often used when growing marijuana indoors. This was enough to get

---

[31]   United Nations (2000) 2-C, pp. 44–47.
[32]   Principle IV.

a Federal warrant to search the house, where the agents did indeed find marijuana growing.[33]

These cases suggest that in the USA, satellite imagery taken in visible light may be deemed noninvasive, but that infrared imagery might be. The courts would have to decide whether infrared imagery is in general public use, such as used by weather forecasters. Until then, infrared satellite imagery is probably out of bounds as a tool for monitoring companies and individuals in the USA without a warrant.

Generally speaking, satellite images alone are not enough to convince a court. First of all, as with any digital image, there is the possibility that the image has been "Photoshopped" – referring to the popular Adobe software for massaging digital photos on a computer. The same software that removes red eyes from a family photo can remove, insert or change pretty much any aspect of an image. Satellite images are usually manipulated on a computer to highlight topics or regions of interest. Figures 56, 57 and 58, appearing and discussed in Chapter 4, show some of the more complicated massaging that is done to help analysts identify what they are looking for. Judges have indeed thrown satellite images out of court because of this. The defendant in a 2001 UK court case argued that the image of his field had been misinterpreted – the prosecution said that his field contained no linseed and thus his claim for a linseed support grant was fraudulent. His defense persuaded the court that the linseed was relatively sparse and had simply been missed by the satellite imaging technique – the prosecution quickly sent a field inspection team to the farm and discovered that the farmer was correct, forcing them to drop the case.[34]

If satellite images are combined with ground inspections, and if the computer massaging they received is carefully documented, courts are generally ready to convict. So, if you are growing something illegal in your backyard, better cover it up every time a surveillance satellite comes over the horizon. If there were only a few such satellites, you could use a website like *www.heavens-above.com* to determine when they would be flying close by. However, the large and increasing number of imaging satellites is such that this is not really a practical plan of action.

The UK linseed farmer was able to argue in court against the evidence from satellites. However, a French businessman was not so lucky in 1997 when his fax messages were picked up by America's ECHELON satellite eavesdropping program – about which more in Chapter 9. As related by James Bamford in his bestselling book about ECHELON,[35] the Frenchman's fax to an Iranian defense official was intercepted by Britain's GCHQ security agency and gave rise to accusations that the French company was selling missile engines to Iran. GCHQ shared the information with its American counterpart, the National Security Agency (NSA), and the issue was rapidly escalated to the political level. America had just persuaded China to stop supplying its C-802 anti-ship missile to Iran; the intercepted fax suggested that the

---

[33] Petras (2005).
[34] Purdy (2006).
[35] Bamford (2001) pp. 411–422.

French company would supply engines that would allow Iran to build its own anti-ship missiles.

The intercepted fax was shared with the CIA, the Commerce Department and Customs in the USA plus similar organizations in Britain. The Frenchman's name was soon on a variety of "watch lists", although, of course, he was not aware of any of this. The USA eventually issued an official complaint to France demanding that the sale of the engines be stopped. The French company claimed that the sale was of electrical generators that could not be adapted to become engines. French export inspectors opened the crates that were waiting to be shipped to Iran and confirmed that they contained generators.

The NSA reanalyzed the evidence and concluded that the fax and other intercepted communications were more ambiguous than at first sight. They admitted that the evidence did not warrant the official complaint, although it would have been a trigger for seeking additional evidence.

So, where did this leave the unfairly accused French businessman? His name was now in the computers of law-enforcement organizations around the world and who knows with whom they shared this information. After NSA reconsidered its analysis and effectively cleared the Frenchman, did they share *that* with these other organizations? Will he be stopped the next time he enters the USA, or even arrested? How many other people are on watch lists based on ambiguous and dubious evidence from satellites? The issue is illustrated by the case of the Malaysian Ms Rahinah Ibrahim, who earned her PhD at Stanford University but has been placed on a watch list that prevents her entering the USA. Having already received $225,000 in settlement from two agencies, she is pursuing her claims against the Federal Government – from outside the country.[36]

There is often a tension between privacy and safety. By tracking your cell phone, the phone operator can advise your family, friends or emergency authorities where you are. Communications by satellite can extend that tracking worldwide. One manufacturer of a satellite tracker advertises "near real-time alerts to designated authorities the moment an employee or asset moves outside of their designated or planned route" and "a dedicated panic button provides fast worldwide E-911 alerting". The advertising emphasizes the ability of companies to track their employees especially in remote or high-risk areas. Steve Edgett, who runs the British-based EMS Global Tracking, says "there is a growing body of regulations in what is called 'corporate duty of care' for lone workers in remote places". He says that failure of an employer to provide this type of tracking may infringe the employer's duty of care and may result in higher insurance premiums.[37] The location of the person being tracked is, of course, determined by a GPS receiver in the satellite tracker, as described in Chapter 6.

---

[36]   McEntire (2010).

[37]   Inmarsat (2010) pp. 64–65, and de Selding (2010b).

COUNTER-MEASURES

How about spoofing the satellites? The military do this sort of thing from time to time, referring to it as "counter-measures", camouflaging missile sites or dotting cardboard tanks and aircraft around the landscape, even putting electrical heaters inside them to make them look active.[38] Or perhaps go a bit further and dazzle the satellite so that the image is over-exposed. China is said to have aimed a powerful laser at an American spy satellite a few years ago, although it is not thought to have done any damage – see Chapter 8.[39] Mirrors to reflect sunlight might achieve a similar effect on satellites taking optical images and a flat metal surface could do the same for satellites taking radar images. Such a scheme may well be legal but, of course, it would attract attention – the image will show a very bright spot even if it doesn't indicate what is causing it.

The UN Principles mentioned above confirm the "freedom of exploration and use of outer space",[40] which is usually interpreted as calling for satellites to be allowed to orbit freely. However, the phrase is sufficiently woolly to mean whatever a clever lawyer wants it to mean. It could mean that satellites in outer space must be allowed to undertake their duties freely, whatever those duties are. But, looking at the earth is hardly *exploring outer space*, even if it is arguably a *use of outer space*.

Interference with surveillance satellites, even by deception, is explicitly forbidden in the arms limitation treaties signed by the USA and the Soviet Union. In the first Strategic Arms Limitation Treaty (SALT-I), each country "undertakes not to interfere with the national technical means of verification of the other Party".[41] The phrase "national technical means of verification" includes spy satellites, and it was interpreted by both sides as meaning that they mustn't cover up their missile silos or submarine bays, thereby preventing the other side from counting and analyzing the missiles and submarines. One legal implication of this is that if another Treaty fails to mention deception, then the tactic is implicitly allowed.

---

[38] Norris (2007) p. 107, and Wright *et al.* 2005.
[39] Covault (2007d).
[40] Principle IV.
[41] Article V.2 of the Treaty, available to download from several websites, such as that of the Federation of American Scientists, *www.fas.org/nuke/control/*.

2

Weather satellites

FORECASTING THE WEATHER

The arrival of the telegraph in North America in the mid 1800s led to the invention of weather forecasting as we know it. Not the "red sky at night shepherd's or sailor's delight" type of forecast, but a recognition that weather conditions are transferred from one place to the next by the winds. Cleveland Abbe is usually cited as the first modern weather observer, initiating a network of weather stations linked by telegraph in 1869. The success of this initiative led to the creation of the national weather service in 1870. Run by the US Army Signal Corps, Cleveland Abbe was appointed its chief scientist.

Today, satellites show us that the weather on the far side of the world affects our weather in 4 or 5 days' time. Sequences of pictures from satellites high above North America and the Atlantic Ocean show cloud formations forming in the Pacific Ocean and traveling all the way across Canada before eventually reaching Europe. This sequence doesn't happen all of the time – that would make forecasting too easy – but it does happen sufficiently often to demonstrate the linkage in a general sense of Europe's weather today to that in the Pacific Ocean off America's West Coast several days before.

This inter-continental linkage of weather systems was recognized during World War II when it was realized that the so-called "fire-balloons" landing in the USA and Canada had traveled in about 3 days all the way across the Pacific Ocean from Japan. The balloons were carried along at above 30,000 ft (9 km) by what we now call the Jet Stream.

Before the age of satellites, weather bureaux collected data from weather stations and ships, buoys and balloons – the balloons carried weather instruments such as a thermometer and barometer and a radio to send the data collected as they rose through the atmosphere back to the ground. The data tended to come from wealthy nations, to be very sparse over the oceans and poor countries, and to be particularly sparse in the southern hemisphere and in both polar regions. Satellites have made the

P. Norris, *Watching Earth from Space,* Springer Praxis Books,
DOI 10.1007/978-1-4419-6938-5_2, © Springer Science+Business Media, LLC 2010

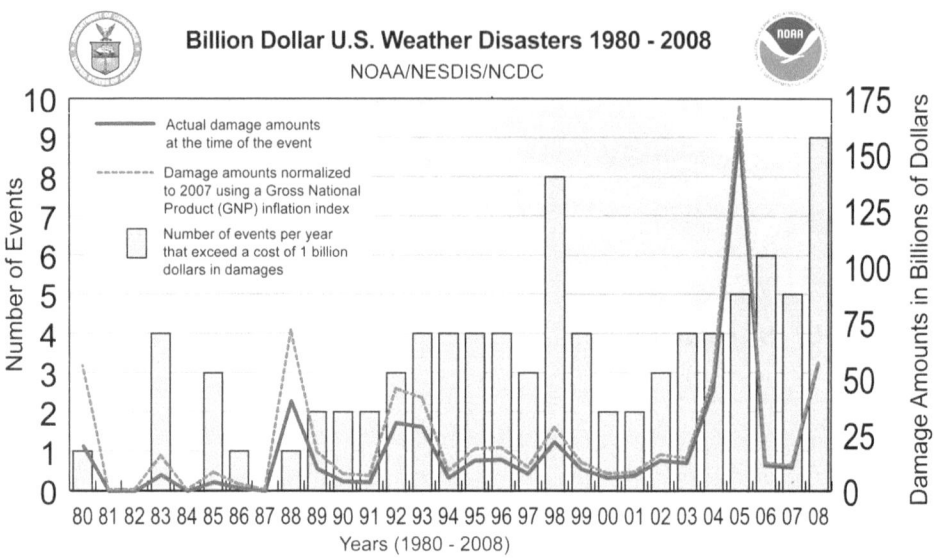

Figure 8. Economic losses caused by US weather disasters. Credit: NOAA/NESDIS/ NCDC.

location of the data more democratic – satellites don't distinguish between rich and poor countries, or between land and sea. Satellites spot the clouds – at least the tops of the clouds – and can measure sea state (the height and direction of waves), sea surface temperature, wind speed and direction, humidity and some of the chemical constituents of the air such as ozone. Some of these satellite measurements can only be made under clear skies and in daytime, but many are possible whatever the weather or the time of day.

Satellites suffer from the fact that they are hundreds, or in some cases thousands, of miles above the earth. Being on the spot with your thermometer, your wind vane, your humidity reader, your barometer and so on is likely to provide more accurate measurements than those taken by a distant satellite. And satellites are expensive – the costs range from tens of millions of dollars to a billion or more in the case of the most sophisticated satellites. But ships are expensive – millions of dollars – and you would need tens of thousands of them to cover all the oceans. So weather forecasters have come to realize that satellites are the cost-effective option, the way to get the best bang for your buck. As a consequence, the British Met Office says that over 70% of the data they use in their most accurate weather forecasts come from satellites – and the proportion is growing.

The money is deemed well spent. Dr Jack Hayes, a top official in the US weather agency, claims that a third of the US economy is sensitive to weather and climate (see Figure 8). He goes on to list the results of bad weather, including 7,400 deaths and 600,000 injuries each year and two-thirds of air traffic delays. He reckons

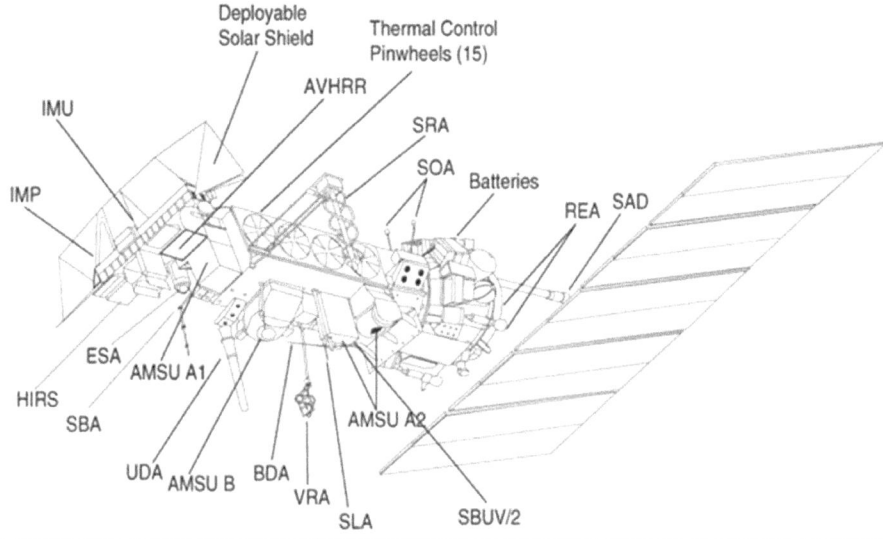

Figure 9. Schematic diagram of the current US low-orbit weather satellite. Credit: NOAA.

LEGEND			
AMSU	Advanced Microwave Sounding Unit	SAD	Solar Array Drive
AVHRR	Advanced Very High Resolution Radiometer	SAR	Search And Rescue
BDA	Beacon Command Antenna Vhf	SBA	S-Band Antenna
ESA	Earth Sensor Assembly	SBUV	Solar Backscatter Ultraviolet Spectrometer
HIRS	High Resolution Infrared Radiation Sounder	SLA	Search And Rescue L-Band Antenna
IMP	Instrument Mounting Platform	SOA	S-Band Omni-directional Antenna
IMU	Inertial Measurement Unit	SRA	Search And Rescue Receiver/Real-Time Antenna
MHS	Microwave Humidity Sensor	UDA	Ultra-High Frequency Data Collection System Antenna
REA	Reaction Engine Assembly	VRA	Very High Frequency Real-Time Antenna

that improved forecasts offer a saving of $18 billion in the air transport sector alone.[1] A year earlier, he stated that much of the improvement in recent weather forecasts was due to weather satellites. "A three day forecast with satellites is as good as a one day forecast without satellites," he claimed. His National Oceanic & Atmospheric Administration (NOAA) colleague, Mary Glackin, told a House of Representatives Subcommittee in June 2009 that NOAA could not produce useful 4 and 5-day hurricane track forecasts if there was a gap in satellite coverage of 6 months.[2]

The drawing of a modern weather satellite in Figure 9 suggests how complex they are. Costing $564 to build and launch, the NOAA-19 satellite was in the news long before it reached the launch site. The two most prominent features in the drawing are the solar panel on the right and the sunshade (called "solar shield" in the drawing) on the left. The 6-m (20-ft)-long solar panel is covered with solar cells on the side facing the sun and provides more than 800 W of electrical power. The sunshade protects the cameras from direct sunlight much as you might shade the lens of your

[1]   Hayes (2009).
[2]   Hayes (2008), and Canan (2009).

Figure 10. NOAA N-Prime (later NOAA-19) on the floor at Lockheed Martin's factory in Silicon Valley, California. Credit: NASA.

camera with the flat of your hand when taking photos in the sun. The cameras and other instruments themselves are relatively inconspicuous. The main camera is the AVHRR, which is located under the sunshade on the left. The other instruments that measure the temperature, humidity and constituents of the atmosphere are placed along the underside of the $4\frac{1}{4}$-m (14-ft)-long body. The right-hand end of the satellite contains batteries, small rocket motors to control the orbit and pointing direction, fuel tanks and other "housekeeping" items. The overall weight of the satellite including $\frac{3}{4}$ ton of fuel is $2\frac{1}{4}$ tons.

NOAA-19 made the news in 2003 for reasons illustrated in Figure 10. Following a change of shift, the Lockheed Martin team that was building the satellite started to lower it from the vertical to the horizontal so that they could gain access to the Microwave Humidity Sounder. The previous shift hadn't inserted the 24 bolts that held the satellite to the gantry, assuming the next shift would, and vice versa – and the paperwork signed off at start and end of shifts suggested that the bolts were in place. Newton's laws of gravity apply in Silicon Valley just like everywhere else and the 1-ton satellite smashed to the floor, thankfully without injuring anyone.

Nearly 6 years and $250 million later, NOAA-19 had been repaired and was ready to be launched, as illustrated in Figure 11. This figure gives a good impression of the size of the satellite compared with the engineers at its base. The bolts on this occasion have been put in place to prevent it falling as it is raised from the horizontal to the vertical. Eventually, it was successfully placed in orbit on February 6th 2009.

Table 1 lists all current low-orbiting operational weather satellites. There are, in

Table 1. Low-orbit weather satellites as of July 2010
(satellites to be launched in 2010 in italics).

Country	Satellites (launch dates)	Parameters measured
USA	NOAA-15 (1998), NOAA-16 (2000), NOAA-18 (2005), NOAA-19 (2009), DMSP-F14* (1997), DMSP-F15 (1999), DMSP-F16 (2003), DMSP-F17 (2006), DMSP-F18 (2009)	Multi-spectral low-resolution imaging, atmospheric temperature and moisture content, stratospheric ozone, aerosols
Europe	Metop-1 (2006)	Similar to USA plus surface winds
China	FY-1D (2002), FY-3A (2008), *FY-3B (2010)*	Similar to USA plus ocean color imager
Russia	Meteor M-N1 (2009), *Meteor M-N2 (2010)*	Prototype – similar to USA plus low-resolution imaging radar for ice monitoring

addition, more than 30 satellites whose data assist weather researchers to understand the subject better but are not routinely used in weather forecasting.[3]

Many countries have weather-forecasting bureaux and you might wonder at the economics of having separate (and separately funded) forecasters in neighboring small countries. Fortunately, this fragmentation has been mainly avoided when it comes to weather satellites. Through the World Meteorological Organisation (an agency of the UN), countries have agreed to share their weather satellite data with each other. They have also divided up responsibility for providing weather satellites at an altitude of 36,000 km around the equator – the altitude at which the satellite is moving at the same speed as the earth is rotating below, thus appearing stationary in the sky above the equator (the satellite is said to be in a "geostationary" orbit). A geostationary satellite is so high in the sky that it can see almost all the way to the poles – the image becomes foreshortened and thus hard to interpret the further away from the equator you go, but data out to 50° latitude is fine, which covers most of the inhabited areas of the globe. The USA has committed to providing two such satellites: GOES West, covering the eastern Pacific from beyond Hawaii to the central part of North America, and GOES East, covering from there to the eastern Atlantic near Africa;[4] Europe (through the Eumetsat organization), China, India, Japan, Korea and Russia also provide similar satellites, as listed in Table 2.

Besides the geostationary satellites 36,000 km out in space, the USA, Europe, China and Russia also provide weather satellites that are at an altitude of about 850 km (see Table 1). These have the advantage of being a lot closer to the weather

---

[3] CGMS (2010) pp. 26–28.
[4] Davis (2007).

Figure 11. NOAA-19 being prepared for launch at Vandenberg Air Force Base, California. Credit: NASA/NOAA.

Table 2. Geostationary weather satellites as of July 2010
(satellites to be launched in 2010 in italics).[5]

Country	Longitude	Name	Launched	Channels	Comments
USA	135°W	GOES-11	2000	5	Operational: West
	75°W	GOES-13	2006	5	Operational: East
	105°W	GOES-15	2010	5	back-up
	89.5°W	GOES-14	2009	5	back-up
	60°W	GOES-12	2001	5	South America
Europe	0°	Meteosat-9 (MSG-2)	2005	12	E. Atlantic
	9.5°E	Meteosat-8 (MSG-1)	2002	12	Rapid scan service
	57.5°E	Meteosat-7	1997	3	W. Indian Ocean
	67.5°E	Meteosat-6	1993	3	Back-up
India	74°E	Kalpana-1 (METSAT)	2002	3	
Russia	*76°E*	*Electro-L N1*	*2010*	*10*	*Planned*
India	*82°E*	*INSAT-3D*	*2010*	*6*	*Planned*
China	86.5E	FY-2D	2006	5	
India	93.5°E	Insat-3A	2003	3	
China	123.5°E	FY-2C	2004	5	Back-up
S Korea	128.2°E	COMS-1	2010	5	
China	105°E	FY-2E	2008	5	
Japan	140°E	Himawari-6 (MTSAT-1R)	2005	5	
	145°E	Himawari-7 (MTSAT-2)	2006	5	Stand-by

than the geostationary (GEO) satellites and can get more detail, but they only see each region of the globe for a few minutes before their trajectory takes them over the horizon. These Low Earth Orbiting (LEO) satellites typically see each part of the globe twice a day, so a combination of LEO and GEO weather satellites gives a combination of data that is continuous but less detailed, and occasional but more detailed.

The five countries that originally agreed to provide geostationary weather satellites (USA, Europe, Russia, China and Japan) have now been joined by India's Insat series and South Korea's COMS-1 satellite. You might wonder why South Korea needs to launch a geostationary weather satellite, since its close neighbors China and Japan already do so. Lee Joo-jin, who is the head of South Korea's space agency, KARI, says "The problem with our meteorological system now is that we rely on information provided by Japanese satellites, which provide information

---

[5]   WMO (2010).

updates every 30 minutes. COMS-1 will reduce this gap to 8 to 10 minutes, which will dramatically improve the accuracy of weather reports and put the country in a better position to prevent damage from natural disasters".[6] The cost of over $300 million seems high for the reduced time delay, and the camera and satellite are being bought from the USA and Europe, respectively, so local industry is not benefitting directly. We can be forgiven for surmising that the unstated reason for COMS-1 being deployed is that being dependent on China and (especially) Japan has sometimes been an unpleasant experience for South Korea in the past.

The first weather satellite was the Tiros satellite, launched in 1960, and some of its features remain unchanged in its successors today. The Tiros camera used TV-type cathode ray tube technology, which today is replaced by the sort of digital technology you have in your cell phone or digital camera. The images showed very little detail and that is still the case today, when the smallest features visible in weather images are typically 1 km or more in size. The lack of detail saves on the bandwidth to transmit the images to the ground – just as you pay more the better your broadband internet connection is. Weather satellites broadcast their images to the ground in two formats. One is a format that can be received by cheap and cheerful terminals. Originally, the format was that of the facsimile or fax, which, in recent years, has been replaced by a digital equivalent. The idea is that small organizations and individuals can afford to own a weather satellite terminal.

The second format contains more detail and is that used by the main weather forecasting agencies. Once received on the ground, it is processed to remove defects and to extract weather information. For example, consecutive geostationary images are processed by computer so that the movement of individual clouds can show the wind speeds and direction – an animation of this process can be seen on the Eumetsat website.[7] However, a small change in the camera or the satellite between or during images might be interpreted as a cloud movement, so, first, the computer checks each image by identifying landmarks and distorting the image so that the landmarks are all in their proper places. Then, the computer uses the shape of the clouds to find how they have moved between successive images – typically 15 min or so apart. Finally, a separate infrared (heat-sensing) image of the same scene is consulted to determine the height of each cloud – the warmer the cloud, the lower its altitude. By this combination of geostationary weather satellite and sophisticated computer pattern recognition, weather forecasters get a snapshot of the winds around the globe every few hours.

Figure 12 shows a typical output of this process from the Meteosat satellite located above 0° longitude (directly above the center of the figure). Weather patterns at different levels are evident over land and ocean.[8]

---

[6]   Kim Tong-hyung (2010).

[7]   *www.eumetsat.int/Home/Main/Access_to_Data/Meteosat_Meteorological_Products/Product_List/SP_1119537341561?l=en.*

[8]   The computer system to extract wind and other weather information from Meteosat images was developed by the team I work with at Logica plc.

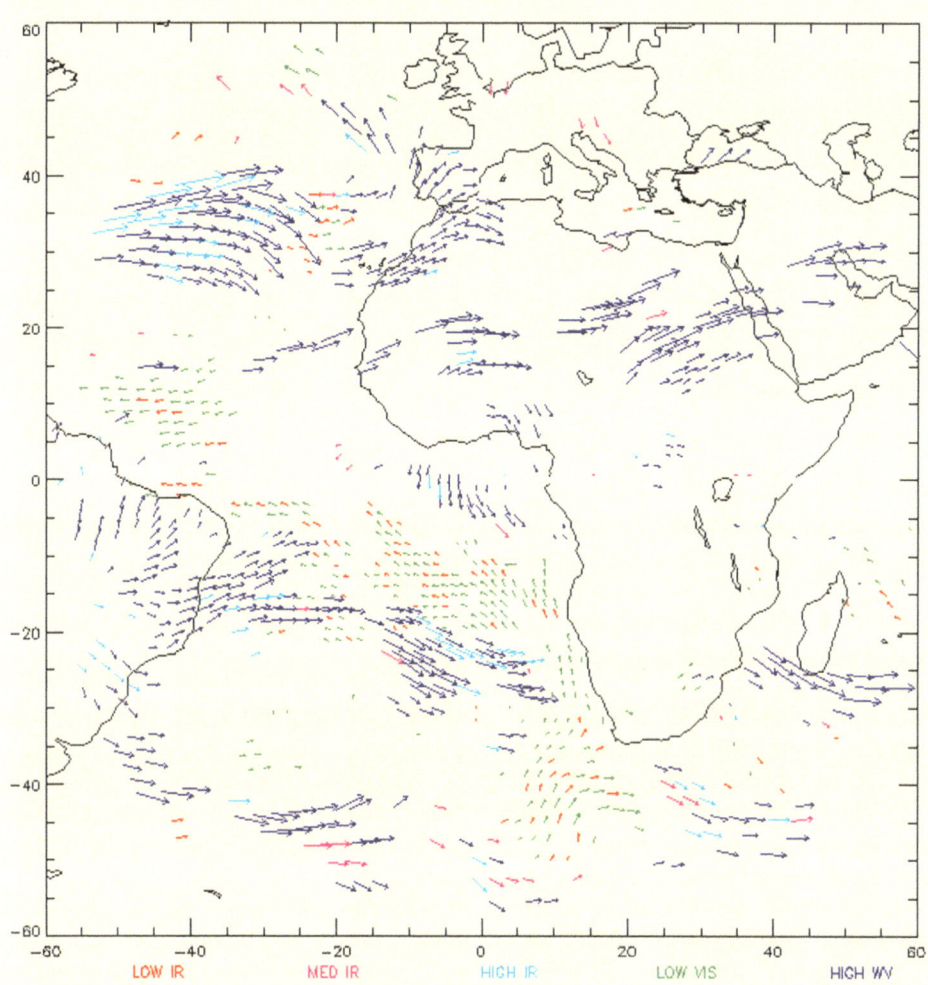

Figure 12. Wind directions (arrow) and speeds (length of arrow) at high (light and dark blue), medium (pink) and low (red and green) altitudes computed from three Meteosat images on April 16th 1997 using data from three channels: visible (green), infrared (red, pink and light blue) and water vapor (dark blue). Credit: Eumetsat.

The list of current geostationary weather satellites in Table 2 includes the number of "channels". The equipment on each satellite takes several images at a time, each in a different color (effectively, each color has its own separate camera). Most of the colors are in the infrared in order to pick out features in the sea or cloud, so the word "color" isn't really appropriate, since infrared is invisible to the human eye. The satellite people speak about bands or channels instead of colors. Europe's two most recent satellites, Meteosat-8 and -9 (previously called MSG-1 and -2), take images in 12 channels, three of which are in the visible part of the spectrum. The other nine

Figure 13. Composite of geostationary images on March 25th 2010 at 06:00 (GMT) with latitude–longitude lines and political boundaries added by computer.

channels are in the infrared, each chosen to detect clouds or humidity at a particular altitude – their use for detecting volcanic ash will be mentioned in the next chapter.

Figure 13 shows how imagery from several geostationary satellites can be spliced together to give a snapshot of the world's weather. The image is in the infrared, where black means hot and white means cold. The time of this particular composite image is 06:00 GMT, so it's early morning in the UK (the *Greenwich* in Greenwich Mean Time, GMT, is a suburb of London). Thus, India and Australia are experiencing the full heat of the day and appear black, whereas it is still nighttime in Europe and America. The whiter the clouds, the colder they are – and cold means high. The light-gray clouds are closer to the surface. Weather patterns stand out clearly, such as a frontal system over the south-east of the USA out into the Atlantic and a circular storm to the east of the Philippines.

In principle, images from a LEO satellite can also be used to spot cloud movements and thus wind speeds. At the equator, there will be many hours between LEO images, in which time the clouds have changed shape or disappeared. However, near the poles, these orbits tend to converge – the satellite orbits from pole to pole so that the earth has turned beneath it each time it gets back to the equator. So, a LEO satellite might get successive images of a polar region after about 90 min (the duration of a LEO orbit), which is fine for tracking individual clouds. As noted above, geostationary satellites give little or no information about the polar regions, hence the usefulness of the LEO data.

Another source of wind information is to measure the direction and height of waves on the sea surface. Several low-orbiting satellites carry instruments that measure this information, but Europe's Metop-1 is currently the only operational weather satellite that does so.

Figure 14 illustrates how satellites pick out weather patterns that are smaller than continental scale but still too large to be seen as a whole by aircraft- or balloon-borne sensors. Taken by NASA's Terra satellite on March 20th 2010, the lower half of this natural color image shows a sandstorm over eastern China that has swept in

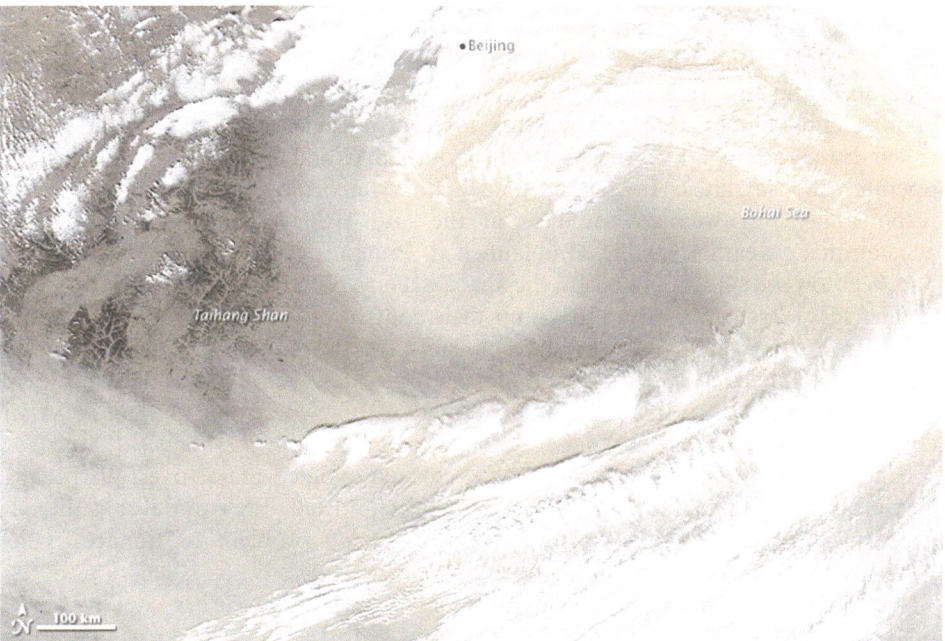

Figure 14. Sandstorm in China illustrating a "comma"-shaped wind pattern captured by the MODIS instrument on NASA's Terra satellite (see text for explanation). The scale is shown in the lower left. Credit: NASA's Earth Observatory.

from the dry dust plateaus to the north and west. NASA's website[9] explains that the sandstorm "wraps around the right-hand side in a comma shape that terminates in a large ball of dust near image center. This pattern is consistent with the passing of a cold weather front bearing a strong area of low pressure at the surface. These weather systems, known as mid-latitude cyclones, are often associated with giant comma-shaped clouds that reveal how air from a very wide area gets drawn in toward the low-pressure heart of the storm". The sand degrades the air quality in China's cities and obscures landmarks to the point at which we cannot see where the land ends and the sea to the right begins. Only the mountains to the lower left are recognizable – the NASA annotator has identified the Taihang Shang (mountains), which mark the easterly edge of the North China plateau.

American weather satellites are owned and operated by the National Oceanic and Atmospheric Administration (NOAA). It might seem odd that NASA doesn't take care of them – after all, isn't it NASA's job to run non-military satellites for the government? NASA did develop the early weather satellites, but it soon became clear that NASA's culture of seeking to always improve the technology was at odds with providing a public service at the lowest possible cost. Why improve something if it does the job and it will cost more to change?

---

[9]   *http://earthobservatory.nasa.gov/NaturalHazards/view.php?id=43207*.

This tension between the space agency and the weather-forecasting agency has been repeated around the world. In Europe, the first weather satellite, Meteosat, was developed and operated by Europe's NASA, the European Space Agency (ESA). Almost immediately, Europe's weather forecast agencies created a special agency called Eumetsat to take over Meteosat from ESA, and Eumetsat now owns and operates the satellites. In Japan, the Japan Meteorological Agency took over responsibility for weather satellites in the late 1990s when the first MTSAT geostationary weather satellite was launched. Previously, Japan's weather satellites had been owned and operated by the space agency NASDA (now called JAXA).

The space agencies haven't given up their control of weather satellites without a fight. In the USA, NASA is still responsible for developing new generations of weather satellites that NOAA then builds and operates. This arrangement has led to budget and schedule overruns as NASA tinkers with new technologies and NOAA is committed to providing the funds.

Europe has a similar situation but has (so far) avoided major overruns because Eumetsat gives a fixed amount of funds to ESA and tells ESA to develop the new generation system – any overruns are then ESA's problem to finance, which motivates ESA to prevent them happening.

The US Department of Defense (DoD) has its own LEO weather satellites, called the Defense Meteorological Satellites Program (DMSP). These 800-kg satellites take broadly similar measurements to the 1.4-ton NOAA satellites but have some encryption features to allow US forces to receive the data on the battlefield while denying it to the enemy. The similarity between the two satellites led President Clinton to decide in 1994 that the next generation of both would be common – called NPOESS[10] (pronounced "en-pose"). That development was run by a project team reporting to DoD, NOAA and NASA, and the result of this confusing management chain has been delays, budget overruns, Congressional hearings, changes of contractors, but no satellites! Initially expected to produce six satellites, each carrying 10 sensors, over a 20-year period, NPOESS had a budget of $8.4 billion. By 2006, the budget had gone up to $12.6 billion but the number of satellites was down to four and the number of sensors to seven. The first interim satellite was supposed to be in orbit by 2006, but by the time the program was cancelled in February 2010, it was still a year and a half away from launch. The cancellation came even after the Obama Administration pumped an extra $100 million into its 2010 budget. A Senate Appropriations Committee report called for the Administration "to disengage from its autopilot management style" and start taking responsible decisions – the cancellation followed. The US military and civilian weather satellites will now remain separate.

The DoD prudently still has two of its current satellites on order that will now be launched in 2012 and 2014, giving it time to plan how best to proceed for the long term.

---

[10]   NPOESS = National Polar Orbiting Environmental Satellite System.

Figure 15. NOAA Administrator Dr Jane Lubchenco claims that splitting DoD and NOAA weather satellite programs makes sense, but sees no inconsistency in NASA and NOAA programs remaining tightly intertwined. Credit: NASA photo by Bill Ingalls.

Unlike DoD, NOAA has no further versions of its existing satellites on order – the last was launched in early 2009. NOAA is left with the interim NPOESS satellite now due to be launched in 2011. The plan is that the first of the next-generation civilian NOAA satellites will be launched in 2014, funded by NOAA and procured by NASA, although past experience suggests that this timetable is optimistic. Commenting on the cancellation of the joint civil–military program, NOAA Administrator Jane Lubchenco (Figure 15) put a brave face on the debacle, saying that "this partitioning will enable both partners to do what they do well".[11]

The Defense Department's weather satellite program has been successful for over 30 years. The NOAA–NASA civilian program is another matter. Inside the Defense Department, the weather satellite program was for many years run by the Strategic Air Command, whose mentality was operational rather than research. The weather satellites were designed to be fit for purpose, and to provide a day-in, day-out regular service. By contrast, the US spy satellites of the Cold War era were the responsibility of the Air Force Space Command, where pushing the boundary (in the technology sense) was integral to the culture. The spy satellites were often late and over budget, but they were technically very advanced.[12]

The US civilian weather satellites have tried to find a compromise between these two cultures, with NOAA as the end customer providing the regular service

---

[11]   Brinton (2010g), Brinton (2010b), and Canan (2009).
[12]   Arnold (2008).

Figure 16. Senior NOAA Official Dr Jack Hayes describes going from research to operations as "bridging the valley of death".[13] Credit: NOAA.

mentality and NASA as the development agency stretching the technology. The results have been mixed. A weakness in the arrangement is that NOAA pays NASA the cost of the development whatever the price; so, if NASA mismanages the development, NOAA pays. This contrasts with the situation in Europe in which the same sort of collaboration calls for Eumetsat (Europe's NOAA) to pay a fixed amount to ESA (Europe's NASA) so that overruns by ESA have to come out of its own budget. The cradle-to-grave cost of the next batch of US geostationary weather satellites, the GOES-R program, ballooned from $6¼ billion to $11 billion before a decision was made to delete one of its new instruments. Even so, the cost estimate is nearly $8 billion and the launch of the first satellite has been put back from 2012 to 2015.[14] NOAA admits to finding it difficult to turn advanced research into operational weather forecasts (Figure 16).

The current batch of US geostationary satellites comprises Geostationary Operational Environmental Satellite (GOES)-13 through GOES-15. Each weighs 3¼ tons when launched, although over 1½ tons of that is fuel, a lot of which is used to reach its final position in orbit, so it starts its useful life weighing about 2¼ tons. Its 8-m-long solar array produces 2 kW of electrical power – 2½ times that of the low-orbit NOAA satellites. The last in this batch, costing about $500 million, was

---

[13]   Iannotta (2008).
[14]   Hayes (2008).

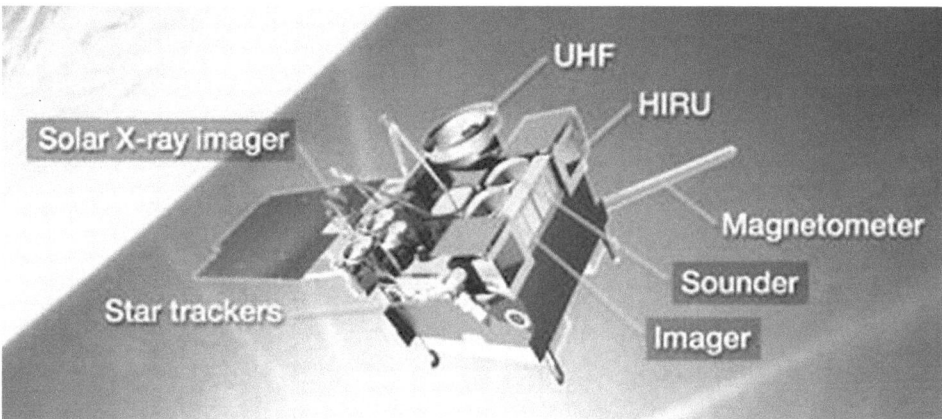

Figure 17. Artist's impression of GOES-13 in orbit. The circular openings of the imager and sounder on the right point to the earth above and to the left in this view. At 36,000-km altitude, the earth should be about 20 times the size of the moon as we see it, so the artist has drawn the earth much too large – almost as if it would fill the sky. The annotation is explained in the text. Credit: NASAscience.

launched in March 2010. There will now be a 5-year wait for the first of the new and improved batch, as mentioned above.

The artist's impression of GOES-13 in Figure 17 illustrates some of the key features of the satellite. UHF is the antenna that receives emergency beacon signals and data from weather stations. HIRU is a gyroscopic unit that, together with the star trackers, keeps the satellite stable and pointing correctly. The solar X-ray imager monitors outbursts of the sun and the magnetometer on the end of the long boom measures the magnetic field of outer space – the long boom avoids it being corrupted by equipment on the satellite. The imager (camera) produces images in five colors (or "channels") of the earth every half hour or so (more frequently if smaller images are taken). Each pixel in its infrared images is 4 km on a side and 1 km on a side in the visible image. The sounder takes a single pixel image in 19 channels, all but one in the infrared, from which we can calculate the temperature, humidity and key constituents of narrow areas of the atmosphere – its pixel size is about 10 times larger (more blurred) than that of the imager – an animation of how a sounder works is available on NASA's website.[15]

There's more than one way to skin a cat, and while Japan's MTSAT satellites look pretty similar to the USA's GOES, Europe's geostationary satellite, Meteosat, looks nothing like either MTSAT or GOES. Figure 18 is an artist's impression of the latest Meteosat in orbit. It weighs 2 tons when launched, of which nearly half is fuel. Unlike the box-shaped GOES in Figure 17, Meteosat is a cylinder about 2 ½ m high and 3 ¼ m in diameter. While GOES stays rock steady as it captures images of the

---

[15]   *www.aqua.nasa.gov/about/instrument_amsu.php.*

Figure 18. Artist's impression of a second-generation Meteosat in orbit; the first of this series, Meteosat-8, was launched in 2002 followed by Meteosat-9 in 2005. Credit: ESA/ Eumetsat.

earth, Meteosat spins at 90 rpm. Its camera stares out through the elliptical window that you can see in Figure 18. Each time the satellite whizzes round, the window is looking at the earth for about 35 milliseconds it then spends 600+ milliseconds staring into space. During those 600 milliseconds, the picture captured during the 35 milliseconds is radioed to earth. The solar cells that provide its 600 W of electrical power are glued to the side of the cylinder rather than being on a dedicated solar panel like GOES or MTSAT.

Spinning is inherently a very stable condition – that is why a spinning top stays upright whereas a stationary one falls over. Meteosat takes advantage of this and avoids the need for the sophisticated stabilization equipment that GOES and MTSAT require. Meteosat's radio antennas are deceptively clever; the "necklace" round the narrow part of the cylinder is a radio antenna that sends radio signals in the direction of earth – electronically cancelling out the spinning movement precisely. Despite looking somewhat old-fashioned, Meteosat has a camera that takes images in 12 channels compared with the five on GOES and MTSAT – Meteosat is therefore the most sophisticated of all the weather satellites in geostationary orbit, at least as concerns taking the most information-rich images. The first of the current generation of Meteosats was launched in 2005 and cost about

$630 million. The series of three satellites plus the cost of rockets to launch them, dedicated ground facilities and operations staff for 12 years came to about $2 billion.

Mind you, Europe has ways of its own to make a muddle of things. The next generation of European weather satellites, the so-called Meteosat Third Generation satellites caused a political furore in the first half of 2010. The early phases went according to the book, with the European Space Agency and Eumetsat collaborating to agree what the satellite would do and obtaining quotes from industry to build the system. The best and cheapest (by about $200 million) industry proposal for the $1¾ billion contract was led by a French company, TAS, with a German company, OHB, earmarked for a large subcontract. The losing bidder had the reverse political pecking order – Astrium Germany as prime contractor, with Astrium France as its supplier. German industry gets about the same number of contracts either way, but the German Government wanted its industry to lead the project and in a fit of pique, vetoed the award of the contract to TAS at the Eumetsat Council.[16] After much German posturing, the French-led bid was accepted but in a way that gave German industry a larger and face-saving role and a small increase in the cost. Another scenario that was considered was for some sort of co-leadership to be set up between France's TAS and Germany's Astrium. This formula was tried for the Galileo satellite navigation program, resulting in costs doubling and schedules slipping by several years as the two rivals fought bitterly over every nut and bolt in the program, and then both companies lost out to a third company when the contract was re-tendered. In the case of Meteosat Third Generation, Europe's politicians avoided a political compromise that ignores commercial realities and fudges industrial responsibility.

The same Meteosat Third Generation project contains another example of Europe's ability to shoot itself in the foot. British companies were barred from competing for the contract mentioned above because two government departments bickered over which of them should pay for the development contract. The Ministry of Defence has to pay for the operational phase at Eumetsat, since it owns the UK Met Office (the national weather service), but feels that the Business Department should pay for the development phase that is managed by the European Space Agency. The two departments failed to reach agreement, so there is no British money in the European Space Agency part of the project, thus barring British industry from it. The Eumetsat part represents three-quarters of the total, so the British money paid into that fund will go towards paying companies in France, Germany and the rest of Europe that have been part of the development phase. British tax payers might wonder whether their civil servants are delivering value for money in this instance.

Meteosat Third Generation will provide sounder information as well as imaging, which the US GOES does already. The idea of a sounder is to measure the

---

[16] de Selding (2010a), de Selding (2010c), personal communications; Portugal also vetoed the award initially but for reasons related to its perilous government finances.

temperature and humidity of the air at various altitudes. A satellite can do this because molecules in the air radiate very faintly in a characteristic way that depends on their temperature. The "color" of the radiation (more correctly, its wavelength or frequency) tells you what height it is coming from. The instrument on the satellite looks downward to measure the radiation at several "colors" – by analogy with a ship, it takes a "sounding", hence the name "sounder" for this type of instrument. The instrument then points slightly to the right, say, and takes another sounding and continues like that to build up a picture of the temperature of the atmosphere below. A computer animation on the NASA website illustrates a typical sounder in action (see *www.aqua.nasa.gov/about/instrument_amsu.php*).

Some "colors" of radiation are affected by humidity in the air, so this information can be used to provide a map of humidity at various heights in the atmosphere. Weather forecasters increasingly use the information from sounders in helping to predict the weather.

Unlike the US GOES, the Europeans will not squeeze the sounder onto the same satellite as the imaging camera, but will fly it on a separate satellite. Over its 20-year lifespan, the Meteosat Third Generation's $4½ billion budget will pay for four imaging satellites and two with sounders. Separating the sounder from the imager means you have to build extra satellites and buy rockets to launch them, but it brings two important benefits. First and most obviously, each satellite is simpler, smaller, cheaper, lighter (thus cheaper to launch) and less likely to run over budget than a complex combined imaging/sounder satellite. The sounder is constantly scanning from side to side, creating jitter in the imager, so a satellite containing both imager and sounder requires special insulating arrangements to reduce this jitter. Second, and perhaps less obviously, if, say, the sounder fails but the imager is still working, you just launch a replacement sounder – vice versa if the imager fails. With the GOES and MTSAT approach in which sounder and imager are on the same satellite, replacing a sounder means launching a whole new expensive imager plus sounder, one of which is not really needed.

Figures 19 and 20 illustrate the power of geostationary satellites to provide continental-scale weather information. Figure 19 is a so-called false color image of North America taken by the US GOES-12 satellite. The image in the visible channel is a black-and-white image, so infrared images are combined with it to create a natural-looking but "false" color image. A new image every half hour enables emergency authorities to track hurricanes as they move across the map. Figure 20 is a black-and-white image taken by Japan's MTSAT-1R satellite, in which the sea and land have been colored in by computer.

Each GOES satellite has an expected life of 10 years and, often, they keep working for longer than that. Occasionally, one fails before it should and leaves a gap in the weather information. One of the advantages of coordinating the satellites of Europe, the USA, Japan, etc. is then evident, in that one of the other countries can lend a satellite to fill in for the failed one for a while. Over the years, the USA has borrowed one from Europe, and Japan and Europe have each borrowed one from the USA. As is evident in Table 2, there appear to be more than enough weather satellites in geostationary orbit at the moment; you need about four to cover the

Figure 19. GOES-12 captured a dying Tropical Storm Danny on the US east coast (top right) and an explosive Tropical Storm (later Hurricane) Jimena on the Mexican west coast (bottom left) on August 29th 2009. Credit: NASA/GOES Project.

whole globe, and there are currently 16 in orbit, although some are only partially working.

Another feature of the coordination is that each satellite relays weather data from weather stations on the sea, on ships and on aircraft, irrespective of the origin of the weather station. This is particularly convenient for airborne stations, since a long-distance aircraft may well move from the coverage zone of one satellite to that of another in the course of a single flight.

The world has therefore got itself a global weather satellite system without the need for a global weather satellite organization. The various weather agencies representing the satellites in Tables 1 and 2 meet from time to time to discuss their plans and to agree how they will coordinate among themselves.[17] The arrangement is hardly perfect, as witnessed by the considerable over-supply of geostationary satellites, but each country decides what it wants to contribute and then just gets on with it. The information from the satellites is shared among the world's weather bureaux free of charge. India is a partial exception to this free-access policy. Until

---

[17]   The group calls itself the Coordination Group for Meteorological Satellites (CGMS); see *http://cgms.wmo.int/CGMS_home.html*.

03:00 UTC 28 MAR 2010   MTSAT JMA

Figure 20. Japan's MTSAT-1R satellite monitors the weather on March 28th 2010 from India on the extreme left to the central Pacific Ocean, and from Antarctica in the south to the Arctic Ocean in the north. Tropical Cyclone Paul is generating winds of up to 90 mph (140 kph) in northern Australia in the lower center.

1999, India encrypted the data sent back by its weather satellites so that only organizations inside India could access it. In 1997, NASA and NOAA signed an agreement with India to allow INSAT data to be accessed by those two US agencies but by no one else, and by 1999, data were being received in the USA. Even then, the data made available to the US agencies are 3 days old, not in real time.[18] In Geneva, the World Meteorological Organisation (WMO), which is an

---

[18]   Chesters (2009).

agency of the UN, keeps a watching brief on all of this, but, in practice, plays no significant role.

GROUND FACILITIES

Speaking of budget overruns – while you might expect some examples in the world of rocket science and satellites, it may come as a surprise to find that the ground facilities are often just as late and over budget.

Take, for example, the software to remove defects from the images provided by geostationary satellites mentioned above. Europe's latest geostationary weather satellite, called Meteosat Second Generation (MSG), needed that kind of software in the control center in Germany. Sadly, the satellite manufacturer that Eumetsat chose to supply that software made a complete mess of it. The contract said that the software was to be delivered a year or so before the launch of the satellite, but it ended up being delivered about 5 years late. Eumetsat decided they couldn't launch the satellite if this particular software wasn't available because you have to remove the defects in the images to get reliable weather information from them. Having delayed the launch by 2 years and incurring tens of millions in payments to the launch company, ESA and the satellite manufacturers (who had to store the satellite and keep their team of experts on stand-by), Eumetsat decided to go ahead, even though the software was still not ready. For the first 2 years after the launch, Eumetsat had to use a stop-gap piece of software that allowed them to produce one set of good images and data a day. Most of the other software for MSG was also delivered late, although not so late as to cause the delay of the launch.

A similar fate befell NOAA's new-generation geostationary satellite system in the 1990s. The satellite was launched as planned, but it was about a year after that before the images and data were of the expected quality due to errors in the defect-removal software.

Happily, some software is delivered on time – the defect-removal software for Japan's MTSAT new-generation geostationary weather satellite worked perfectly first time. And although Europe's MSG suffered from several software snafus, the sophisticated software that automatically extracts wind information (by tracking clouds between images) was on time and working perfectly.[19] The reasons why some suppliers of software mess up while others get it right is a subject for another book – one lesson to learn is that just because a company can build ultra complex and sophisticated satellites doesn't mean they can supply software; as so often in life, it is a case of horses for courses.

---

[19]   The successful MSG and MTSAT systems quoted were both supplied by the team I work with at Logica plc.

HURRICANES, TYPHOONS AND TORNADOES

Satellites have shown themselves particularly useful in alerting communities to the dangers of incoming hurricanes and typhoons.[20] A storm earns this title if its wind speeds exceed 74 mph (119 kph). They start as ordinary storms in the tropical ocean regions, where geostationary satellites can spot their signature spiral shape. If the waters are warm enough, the storm will strengthen and its winds will pick up speed. Satellites allow forecasters to spot the emerging storms, to follow their path and growth, to monitor the temperature of the sea surface and to direct planes and ships to take more detailed measurements if necessary.

Forecasting the path and growth of these big storms is not yet a precise science. One of the factors to consider is the temperature of the sea in the region – the hotter that is, the more likely the storm will grow. Satellite measurements of sea surface temperature are the primary source of this information in most cases.

Tornadoes are a much more transient form of storm, arising and disappearing in hours rather than the days or weeks of hurricanes. The wind speeds in a tornado are phenomenally high, up to 500 kph, hence their unstoppable destructive effect. But they are small in extent compared to a hurricane – less than 100 m wide as opposed to the 100 km or more of a hurricane. Satellites therefore have a tough time spotting tornadoes – the typical geostationary satellite that is constantly watching the USA has a pixel size of a few kilometers – much too big to spot an individual tornado. The best the weather man or woman can do is to spot weather patterns that are likely to spawn tornadoes, but not the tornadoes themselves.

Designs of geostationary satellites that could spot an individual tornado are on the drawing board – but are likely to stay there for quite a while due to their high cost. The military, too, would like a geostationary satellite that could pick out details on the ground and we will discuss the challenges in creating such satellites when we come to discuss military surveillance in Chapter 8.

---

[20]   Hurricane is the name of this type of storm in the Atlantic Ocean, typhoon in East Asia, (severe) tropical cyclone or (severe) tropical storm elsewhere; see *www.aoml.noaa.gov/hrd/ tcfaq/A1.html.*

3

Climate change

HOW HUMANS CHANGE THE EARTH

Mankind has been pumping carbon dioxide into the atmosphere since early humans lit fires for cooking and heating. In those camp fires, carbon in wood combines with oxygen in the air to produce carbon dioxide while releasing lots of heat and light. From about 10,000 years ago, after the last ice age, humans have deliberately burned down trees to clear the land for farming or as part of a tactic for hunting game – flushing them from the cover of the trees.

Carbon dioxide has been increasing slowly in the atmosphere for the past 10,000 years but until about 200 years ago, that may have been from natural causes as much as from human actions. Some scientists argue that the effects of early farmers on the environment were enough to prevent the return of the ice age – the various consequences of human activity raising the temperature to some extent. What is clear is that about 200 years ago, carbon dioxide in the air began increasing rapidly and this is usually linked to the start of the industrial revolution. While it had increased by less than about 7% since the end of the last ice age, namely in 10,000 years, it increased another 7% in 120 years and is currently increasing by 7–8% in 15 years[1] (see Figure 21).

The industrial revolution began the burning of coal in huge quantities to power the furnaces that forged the steel in the foundries and powered the steam engines first in the factories and mines, and then on the railways. Beginning in Great Britain in the late 1700s, the harnessing of coal and steel in this way soon spread across Europe and North America. It resulted in a vast movement of people from countryside to cities, which became grimy and murky from the smoke of the furnaces and fires.

The big question is what effect this increase in carbon dioxide is having on climate. Scientific opinion has concluded over the past 20 years that the result is an

---

[1]   Solomon *et al.* (2007) pp. 23–28.

P. Norris, *Watching Earth from Space,* Springer Praxis Books,
DOI 10.1007/978-1-4419-6938-5_3, © Springer Science+Business Media, LLC 2010

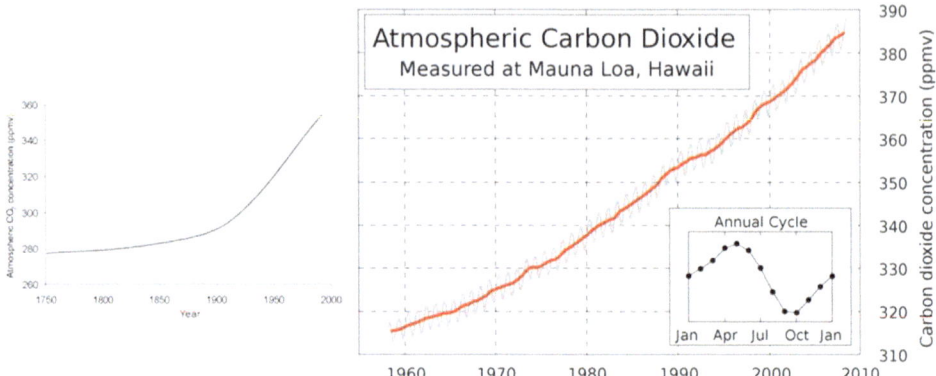

Figure 21. Carbon dioxide content in the atmosphere for the past 250 years (left) and in more detail for the past 50 years (right).

increase in the world's temperature because carbon dioxide acts as a greenhouse gas. The phrase "greenhouse gas" is used to remind people of the way in which a greenhouse keeps warm – the idea is that the glass lets sunlight in, but lets out less of the heat that builds up inside. Similarly, carbon dioxide lets sunlight into the atmosphere but stops heat radiating back into space.

The reason why a greenhouse gets warm seems to be more due to the elimination of drafts than to the sophisticated properties of glass, but, nevertheless, the phrase "greenhouse gas" has stuck as a descriptive way to explain the effects of carbon dioxide in the air.

It would be nice if we could monitor which countries or regions are pumping carbon dioxide into the air. We could then fix quotas for each country or region and use fines or other inducements to ensure compliance. However, carbon dioxide is present in the air naturally and is not easily measured on a country-wide or regional scale. Scientists have had to work out how carbon dioxide and global warming affect the world, and try to monitor each of the effects – sea level rise, temperature rise, retreat of glaciers, melting of arctic ice, etc.

Other chemicals resulting from human activity contribute to global warming such as methane from decay of man-made waste and from animal farming – discussed further below. Carbon dioxide was the first man-made chemical to be recognized as having major climate consequences, but global warming is affected by humans in many different ways.

In order to decide what action is needed and then to check that those actions are working, we need to take good measurements of the earth's environment all the time. Each of the possible effects of global warming needs to be measured – everywhere. The scale of the task is daunting. We have to keep reminding ourselves that "nature doesn't do bail-outs" and press on with trying to monitor what is happening and then take remedial action.

SEA LEVEL

Take sea level, for example. Global warming is thought to raise sea level by a few millimeters per year – 2 mm/year in the 20th century has risen to currently 3 mm/year now.[2] But, each day, even the calmest sea rises and falls by meters because of the tides (except inland seas, such as the Mediterranean, which are less affected by tides). It is not easy to spot changes of millimeters when the effects of tides (and waves) swamp the measurement; the only way is to gather measurements over many years so that the accumulation of "a few millimeters" will become noticeable against the daily tidal ups and downs. If you measure over many years, you have to watch out that the tide gauge (as the measuring instrument is called) doesn't sink slightly as land often does on the coast – a sinking tide gauge will appear to measure a rising sea level. Land rises in some parts of the world, such as in Scandinavia and Canada, as the earth gradually recovers ("rebounds" as the scientists say) from the huge mass of glaciers that covered it during the last ice age. A rising tide gauge will appear to measure a falling sea level. Scientists estimate the amount of rise or fall of the land at each tide gauge and remove that from the sea level figures.

You also have to take these measurements across the globe. Tide gauges are inevitably constrained to be on land, so sea level in vast areas of open ocean goes unmonitored. Satellites can address this issue. The idea is to have an altimeter on the satellite that measures the distance from the satellite to the ground below. If you know the trajectory of the satellite precisely, then you can work out the altitude of the ground below. The key here is that you know the satellite trajectory from Newton's laws, the same laws that allow us to predict eclipses of the sun and moon. Over Mount Everest in the Himalayas, the satellite altimeter will show that the distance to the ground is 8.8 km (29,000 ft) shorter than when over the ocean. Over the ocean, the altimeter measures the distance to the sea below which can be averaged to smooth out the effects of waves.

Satellites circle the globe, so, day after day, year after year, a satellite altimeter measures how far below the satellite the sea surface is. Figure 22 shows the results from satellites developed by the USA and Europe – TOPEX/Poseidon until 2002, the Jason series thereafter. Sea level is shown as currently rising at more than 3 mm/year averaged across the globe.

A global average hides a lot of regional variation. Figure 23 shows the sea level changes for the same period as Figure 22 but on a regional basis. In some areas, sea level is rising at 10 mm/year while in others it is falling at 10 mm/year. The rises outweigh the falls, thus giving the global average rise of 3¼ mm/year.

A satellite altimeter has another big advantage over tide gauges in that just one instrument measures sea level over the whole world. It takes thousands of tide gauges to get close to this and it is difficult to ensure that all of the gauges are accurate.

One of the big challenges with a satellite altimeter is to know the satellite's

---

[2]   Bindoff *et al.* (2007) pp. 408–419.

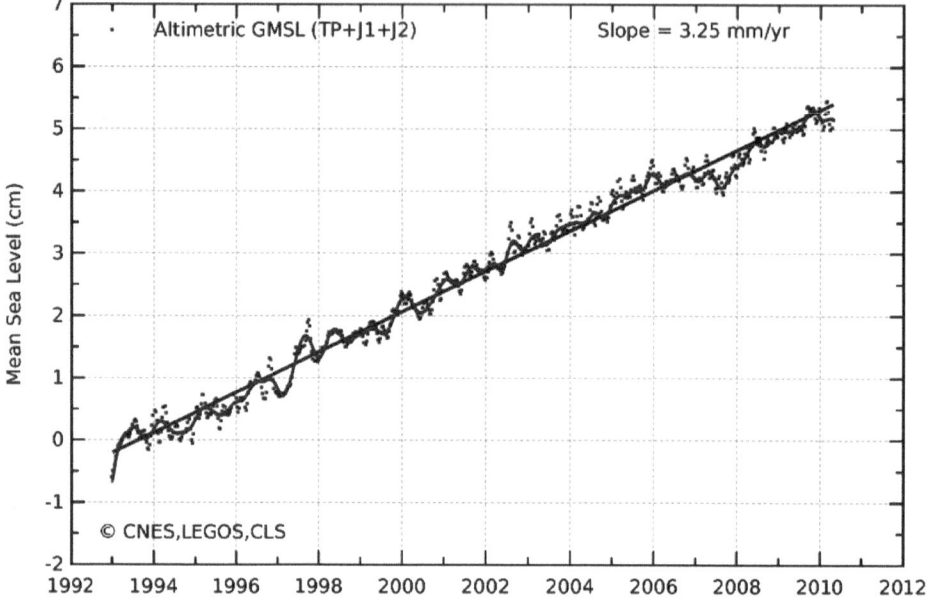

Figure 22. Global mean sea level as measured by the TOPEX/Poseidon, Jason-1 and OSTM/Jason-2 satellites from 1993 to 2009. Each point is a 2-month average; the solid curving line is a 6-month average. Credit: CLS, LEGOS, CNES.

trajectory with an accuracy of a millimeter or better. The best way to check this is to look at the altimeter reading over a fixed point on the earth in two successive passes by the satellite. If the altimeter readings differ by a few millimeters at one of these "crossover points", then our assumption about the trajectory is modified to remove the difference. With information from lots of these crossover points plus tracking data from radars and lasers around the world, the satellite trajectory is computed with the required accuracy.[3]

The importance of monitoring sea level has increased as we discover that it could change much more rapidly than the 3 ¼ mm/year mentioned above. A recent report found that sea level during the recent ice ages went up and down as much as 2 m in a century, which averages out as 20 mm/year or six times faster than the current value.[4]

---

[3]   See, e.g. Scharroo and Visser (1998), Zandbergen *et al.* (1997), and Tapley *et al.* (1994).
[4]   Dorale *et al.* (2010).

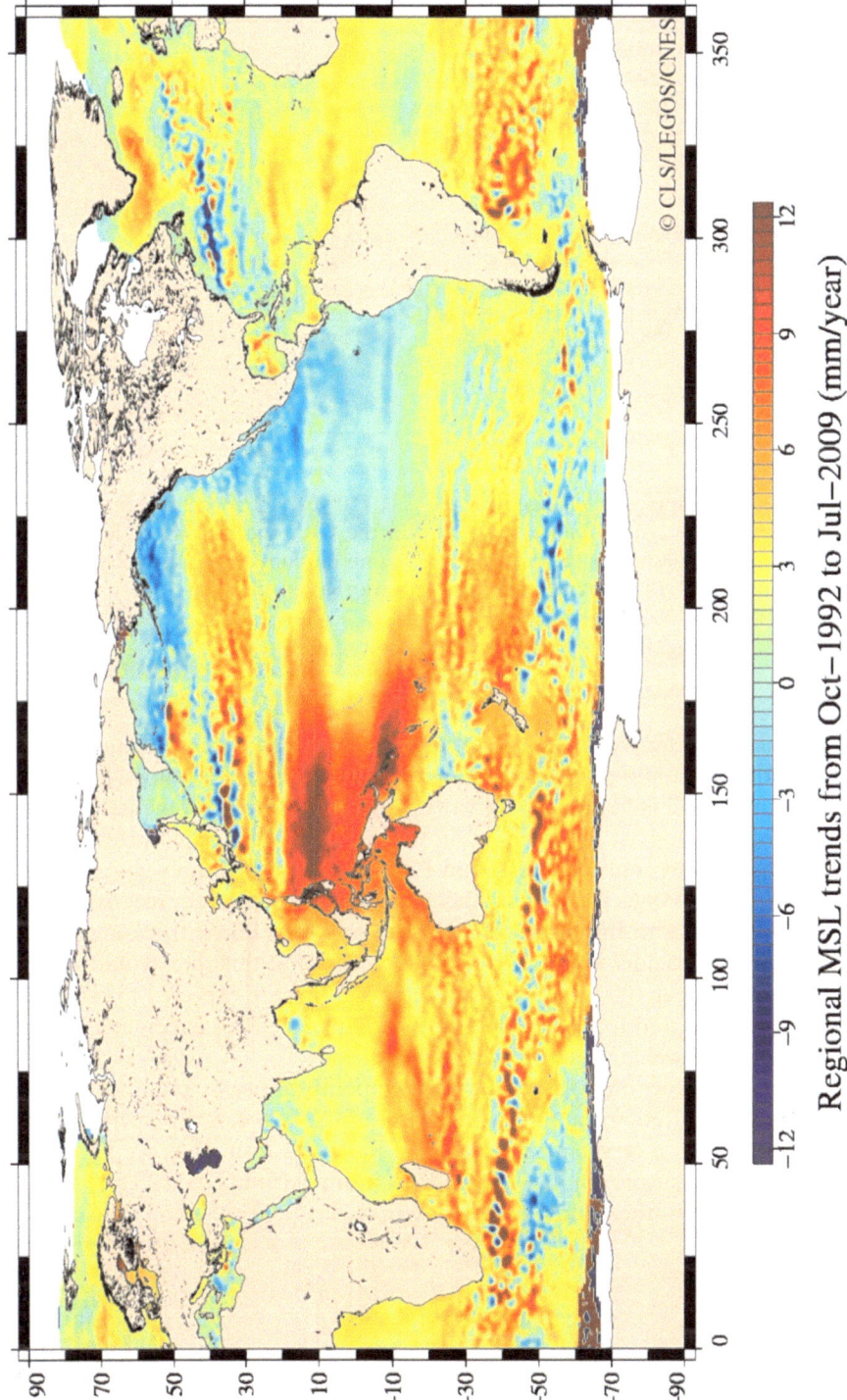

© CLS/LEGOS/CNES

Regional MSL trends from Oct–1992 to Jul–2009 (mm/year)

Figure 23. Regional variation in mean sea level (MSL) measured by satellite altimetry. Credit: CLS, LEGOS, CNES.

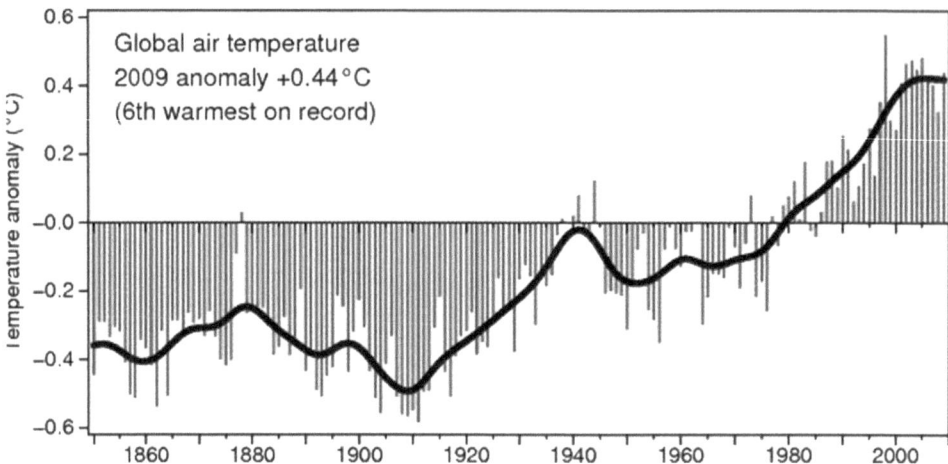

Figure 24. Global temperature change over the past 150 years. This graph comes from the controversial East Anglia University group in the UK, once the world leader in historical climate studies under Professor Hubert Lamb but mired in controversy under Professor Phil Jones.[5] © Copyright 2010, Climatic Research Unit.

GLOBAL WARMING

By and large, temperature varies by tens of degrees from day to night, from one day to the next, and from summer to winter. Thus, the first reaction to being warned of a 1°C rise in temperature is likely to be a yawn. A graph of the average temperature across the globe for the past 150 years is shown in Figure 24 and indicates that the temperature has risen by somewhat less than 1°C since 1860. The Inter-governmental Panel on Climate Change (IPCC) is the most authoritative source of climate information. According to the IPCC, global temperature is currently rising at one-sixth of a degree per decade, which, if it were to continue at this rate, would mean a rise of 1.77°C in the next 100 years.[6] In fact, temperatures are expected to rise faster than the present trend and to have gone up by perhaps 3 or 4°C, depending on how much carbon dioxide we continue to pump into the sky.[7]

Scientists illustrate the importance of a 1°C temperature rise by pointing out that today's temperature differs from that at the height of the last ice age by perhaps only 4°C.[8] So, if a 4°C rise in temperature can move the earth from deep ice age to the current norm, what changes could a few degrees more warming cause?

A temperature rise in the middle of the Sahara desert probably won't have much

[5]   Willis (2010).
[6]   Solomon et al. (2007) pp. 35–37.
[7]   Meehl et al. (2007) p. 749.
[8]   Jansen et al. (2007) p. 435.

effect – the land is already uninhabitable. But if higher temperatures melted the snow in Tibet, a billion people in China, India, Pakistan, Bangladesh and neighboring countries would lose their fresh water. Glaciers in Tibet are considered to be the world's third "pole" after the Antarctic and Arctic regions. Melting snow from the Himalayas and the Tibetan plateau feeds major rivers such as the Yellow and Yangtze in China, the Mekong in Vietnam and Cambodia, the Ganges in India and Bangladesh and the Indus in Pakistan. Recent reports have suggested that the Tibetan glaciers are retreating and disappearing fast, threatening the life-giving water supply for much of South and East Asia. The speed of the glaciers' retreat is a matter of controversy – hence the urgency of getting reliable information.

As so often nowadays, it's not clear whether a recent drought in southwest China (the region nearest to the Himalayas) is due to climate change or is just another in the long series of droughts that affects China from time to time. Reported to be the worst drought to hit China in a century, it has not only left millions of people and livestock without adequate drinking water, but it has also caused electricity "brownouts" due to the reduction in hydroelectric power. Some rationing of power to factories has had to be introduced. About 15% of China's electricity is hydroelectric, making it the world's largest producer of this form of "sustainable" energy, and the amount is due to double by 2020. Rapid development across China has caused deforestation, which, in turn, has reduced groundwater in the southwest. The result will be increased use of coal-fired power stations, already the source of 80% of China's electricity, the cause of much of its air pollution and the source of vast quantities of climate-impacting greenhouse gases.[9]

Figure 25 illustrates how satellites can monitor glaciers in the Himalayas. The growth of lakes on the glaciers is a sign that the glaciers are melting and is readily monitored by satellite. Glaciers in the Andes are also retreating and have lost 20% of their volume since 1975, sometimes with devastating consequences. A piece of the Hualcan glacier fell into a lake in Peru in April 2010, causing a massive wave that swept away houses and factories, killing at least one fisherman. The governor of the region, César Álvarez, blamed climate change. "Because of global warming the glaciers are going to detach and fall on these overflowing lakes," he said.[10]

Measuring global temperature has proved to be difficult and controversial and satellites are only just beginning to help. One difficulty is ensuring that the thermometers don't change over periods of several decades. If it's in a laboratory, you can check it against melting ice and boiling water (0 and 100°C, respectively, by definition), but that's more difficult to do on a mountain top or in a remote valley. One of the problems identified only recently is that cities and towns are hotter than the countryside due to the heat given off by us humans and our activities. Figure 26 shows this vividly in the form of a satellite image of the town and city lights of the USA that illustrate the heat being generated by our urban society.

---

[9] Sheridan (2009), Shiga (2010), and Roberts (2010).
[10] Carroll (2010).

Figure 25. An image from NASA's Terra satellite shows meltwater lakes (blue) in the Himalayas in Bhutan – a tiny state sandwiched between India to the south and China to the north. © Fugro NPA Ltd, *www.fugro-npa.com.*

Figure 26. The USA at night in August 1997 as viewed by US military weather satellites. Credit and ©: NOAA/NGDC, DMSP Digital Archive.

The answer has been to treat with caution temperatures taken in the past by urban thermometers and to move them all to rural sites.

There is also that tricky problem of taking measurements in the middle of the ocean and on top of glaciers. In the past, sailors would toss a bucket over the side of the ship to bring up some water and stick a thermometer into the water. In recent years, it was recognized that wooden buckets give a different temperature from metal or canvas ones, and both give different temperatures from that measured at the inlet for engine room cooling water – this last is the method currently favored.

Satellites use heat-sensing instruments to measure the temperature below. If the satellite is flying over the ocean, it measures the temperature of the "skin" of the sea below. This skin temperature is different from that measured by a tossed bucket or at the cooling water inlet due to being affected directly by sun and wind.

Temperature measurements over ice-covered areas such as Greenland and the Antarctic are sparse even today and non-existent in historical times. Heat-sensing instruments on satellites can measure temperature over the whole of the globe, including polar regions, but the measurements are subtly different from those made with thermometers. For example, some satellite instruments can measure in the dark and through cloud by detecting the minute amounts of microwave radiation (as in a microwave cooker) emitted by the earth below. These microwave measurements are corrupted by passing through the air, especially if it's cloudy, and that affect is difficult to remove accurately. Other satellite instruments measure the thermal signature of the earth below, like the night glasses used by soldiers. These instruments only work in cloud-free conditions and are corrupted by air and cloud. Even comparing satellite and ground temperatures where they coincide in time and place and adjusting other nearby satellite measurements accordingly isn't enough, since the cloud conditions will not be identical.

In summary, the ultra-precise satellite measurements needed to detect temperature changes of a tenth of a degree are only just becoming possible. The good news is that satellites can take measurements across the whole globe and not just where it's easy to place a thermometer or a wind gauge.

ICE

Big changes in the earth's climate in the past have involved ice – glaciers covering huge areas of North America and Europe, mountain glaciers extending far out from the Rockies and the Alps, etc. Arctic ice seems now to be changing fast, with ships able to negotiate both the North-West and the North-East Passages between the Pacific and Atlantic Oceans.

The Arctic ice has changed dramatically since historical records began, most notably during what is known as the "mediaeval warm period" a century or two either side of the year AD 1000. During that period, the Vikings were able to sail to Greenland and colonize it for a few centuries. Viking sagas tell of Eric the Red reaching what is now Canada and this is supported by recent archeological finds. The ice seems to have melted away to the north during this period, allowing the

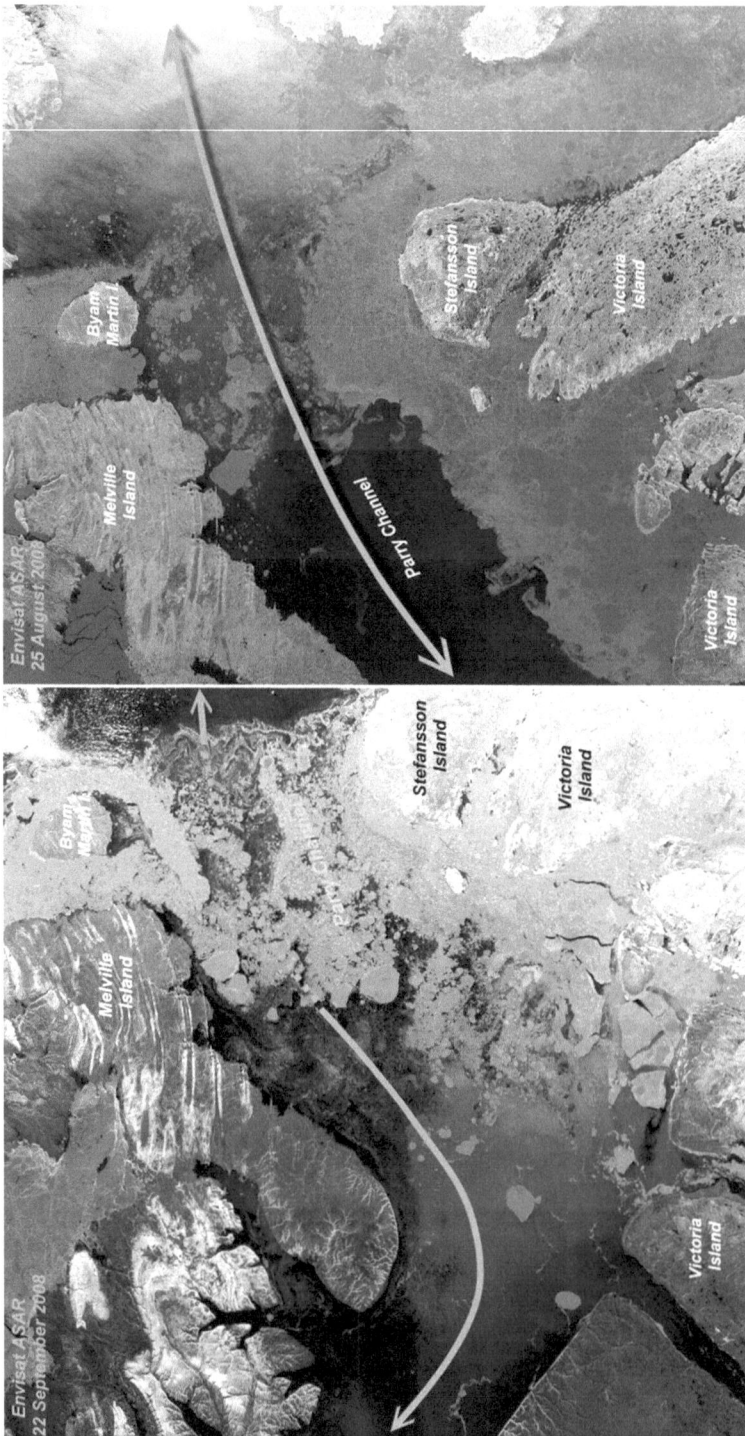

Figure 27. North-West Passage, summer 2008: August 25th (right): open; September 22nd (left): closing. Credit: ESA.

colonists of Greenland to eke out a precarious living. But this happy period came to an end in the 1300s with the return of the ice, causing the Greenland colony to die out.

Before the space age, the only way to reconnoiter the Arctic and Antarctic was to go there in person – typically by ship and sled, and from the 1950s in the Arctic by submarine. The information from these expeditions was very limited in its coverage, giving a fascinating snapshot of one particular place or track but little, if anything, about the polar region as a whole. Satellites now routinely do the opposite. With radar imaging and altimeters that can see at night and through cloud, satellites monitor the overall behavior of the polar regions as illustrated by the following examples.

Figure 27 shows the North-West Passage through the Canadian Arctic archipelago in late summer 2008. The images were taken by the European Space Agency's Envisat satellite using an imaging radar to see through the clouds. The right-hand image shows the Parry Channel, opened in late August, 1,000 km north of the Arctic Circle, and the left-hand image shows sea ice closing it off a month later. The area imaged is about 400 km from north to south. The Irish yacht *Northabout* made the complete circumnavigation through the North-West and then the North-East Passages in 2001–2005 to become the first small vessel to do so since records began, illustrating the significant changes occurring in the Arctic.[11] Figure 28 illustrates the full extent of ice retreat in the Arctic in 2007 using a combination of dozens of imaging radar pictures, again from Envisat. The solid line shows the North-West (left) and North-East (right) Passages in September 2007, with just Vilkitskogo Straits at Russia's most northerly point containing scattered sea ice. Measurements of the reduced thickness of the ice by land-based polar explorers and by soundings from submarines under the ice combined with the irrefutable evidence of the satellite images suggest that the Arctic will be ice-free in summer before long. The extensive melting seen in 2007 did not continue in 2008 and 2009 but the long-term prognosis remains that Arctic ice is gradually receding.[12]

The disappearance of ice in the Arctic Ocean is hard to miss in the satellite images. But it is much more difficult to tell whether the ice in the center of Greenland or Antarctica is rising or falling. We saw above how altimeters can monitor sea level from space and the same is true of ice levels. But what exactly is the altimeter measuring? It bounces a radio signal off the ice and detects the echo – the time between sending the radio signal and detecting the echo tells us how far below the satellite the ice is. But some radio signals penetrate ice at least to some extent. Two altimeters using different radio frequencies will measure different heights. And the

---

[11]   Jarlath Cunnane received the 2005 Blue Water Medal of the Cruising Club of America as builder of the *Northabout* and expedition leader. The skipper, Paddy Barry, was already a Blue Water Medal winner for his previous exploits. Cunnane spent 2002 and 2003 cruising in Alaskan and British Columbian waters, accounting for the long duration of this first east-to-west polar circumnavigation.

[12]   Leake (2010).

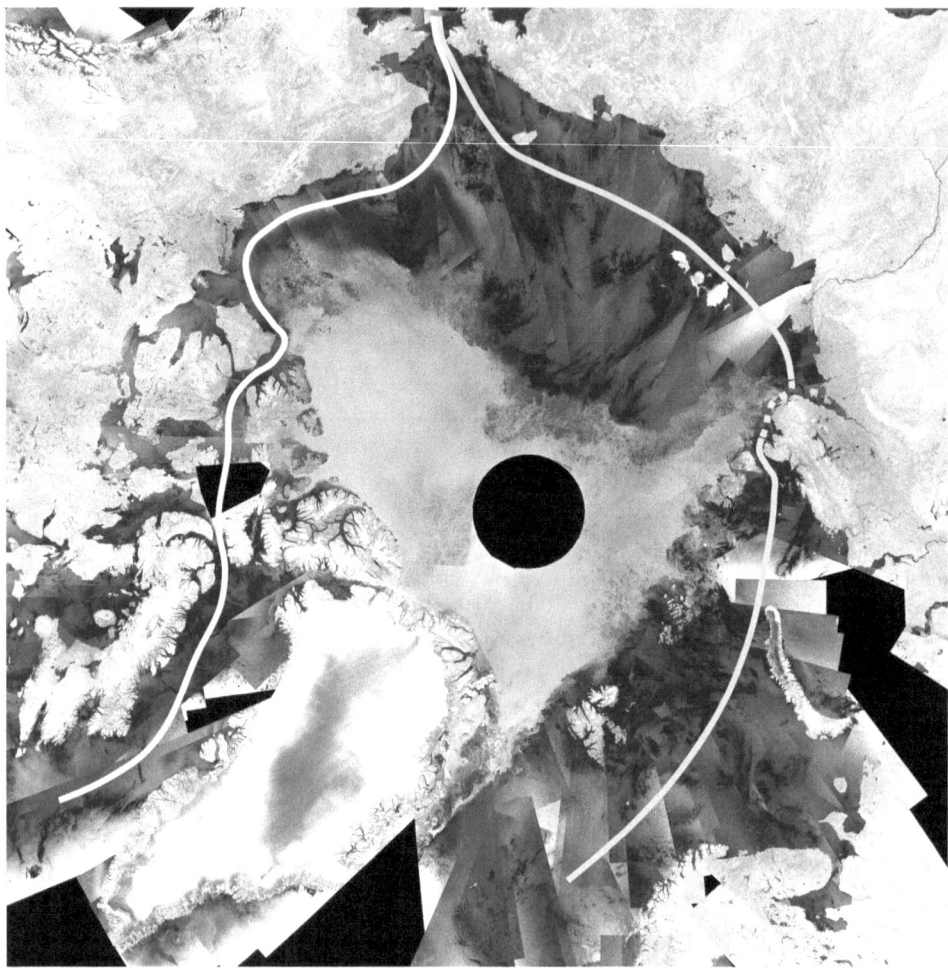

Figure 28. Envisat imaging radar mosaic of the Arctic in September 2007. The central black circle has not been imaged. The solid line indicates the ice-free shipping passage. Credit: ESA.

height also changes depending on whether the ice is wet or dry. These factors have to be taken into account when looking at trends in the ice as shown by satellite altimeters.[13]

Recently, a separate technique has been used to check the results obtained from altimeters. NASA's GRACE satellite measures the earth's gravity, with unprecedented accuracy. Small changes in gravity over a 5-year period allowed the change in the mass of Greenland's glaciers to be measured and gave good agreement with the

---

[13]   Massom and Lubin (2006) pp. 211–216.

Figure 29. Map of northern hemisphere permafrost. North America is on the right, Asia on the left, with the Himalayas and the Tibetan Plateau on the extreme left. Credit: National Snow and Ice Data Center.

results obtained from satellite altimeters. Both methods suggest an annual loss of 200 cubic km of ice weighing 100 billion tons during the first few years of this century.[14] We will meet GRACE again in Chapter 5 in the discussion on freshwater and in Chapter 4 we will discuss the production of 3D maps over ice-free land.

A new satellite with twin altimeters is being designed by the European Space Agency to specifically address the problems of how deep within the ice the altimeter penetrates. Called $CoReH_2O$, its two altimeters will use different radio frequencies and the differences in altitude they measure will indicate the depth of penetration.[15]

As the Arctic warms up, so the permafrost begins to thaw, releasing methane, which, molecule for molecule, has 25 times more heating power in the atmosphere than carbon dioxide. As its name implies, permafrost is permanently frozen and it covers 20% of the earth's land surface, stretching all across Canada, Alaska and Siberia (Figure 29). The parts of the permafrost that are deep underground will stay frozen – and some of it stretches hundreds of meters deep. But almost half of the surface permafrost is within 1 ½ °C of thawing out, so Arctic warming spells wake-up time for this vast source of greenhouse gases.

Just as food rots when the freezer fails at home, when the permafrost soil thaws, microbes consume the dead organic material, producing gases. Melting ice in the permafrost causes the land to subside, creating many new lakes, and these, in turn, speed up the thawing of the surface on the lake bottom. Surveys to gauge the scale of

---

[14]   Chen *et al.* (2009), and *New Scientist* (2009a).
[15]   Clissold (2008); $CoReH_2O$ = Cold Regions Hydrology High-resolution Observatory.

the problem using satellite and aircraft imagery are only just beginning. If present trends continue, a third of the world's greenhouse gas emissions will come from thawing permafrost within a few decades.[16]

Global warming means that the line of permafrost moves closer to the poles. That releases greenhouse gases from the melting permafrost, as just discussed, but it also brings change to the four million people that live in the Arctic and spells trouble for the millions of migratory birds that breed there. Many of these birds currently fly south to the tropics, where forests are disappearing, making their survival more difficult. En route, they rely on a relatively small number of stopover points such as forests, marshlands and tidal pools, where they recover from the extraordinary flights before starting another leg of their journey. These stopover points are shrinking and disappearing at an alarming rate. To cap it all, the northward movement of the permafrost line increases the distances they have to fly and may disrupt the sources of food they depend on. One gloomy but authoritative biologist predicts that "we face the end of [bird] migrations in our lifetime" as a result of this triple whammy impact of climate and global change.[17]

Warmer climate makes the permafrost or tundra greener, but it seems to have the opposite effect on the boreal forests just south of the Arctic Circle. These conifer forests are the world's largest ecosystem and the warming weather hurts them because it is also dryer. Satellite imagery shows that the boreal forests are extending northwards but "browning" because the warmer summers are just too dry.[18]

Iceland is one of the smaller countries in the Arctic region but packs a punch way above its weight when it comes to climate change. Iceland's ice caps have been steadily shrinking for about a century and one consequence is reduced weight of ice on the volcanoes that lie under the island. Spring 2010 showed what can happen when an Icelandic volcano erupts, with air travel across most of Europe grounded for a week. Continued shrinkage of Iceland's ice caps may increase these eruptions in future.

The disastrous effect volcanic ash can have on jet engines was recognized in the 1980s and satellite operators have set up a regular reporting service to provide alerts for the aviation community. One approach is to measure sulfur dioxide, which is one of the toxic and noxious gases given off by an erupting volcano and can be measured by its brightness in ultra-violet light (light that is beyond blue and thus invisible to our eyes) and in certain infrared channels (colors) (see Figure 30). The many channels on Europe's Meteosat geostationary weather satellite allow it to distinguish volcanic ash from clouds by knowing the different brightness of ice clouds, water clouds and ash aerosol in its infrared channels. However, the level of detail is poor, since a single infrared pixel in a Meteosat image covers an area of more than 5 km over northern Europe. A new type of instrument called a Lidar (a radar that transmits light rather than radio waves) should be able to measure the thickness of a

---

[16]   Anthony (2009).
[17]   Weidensaul (1999) pp. 367–370.
[18]   Sturm (2010).

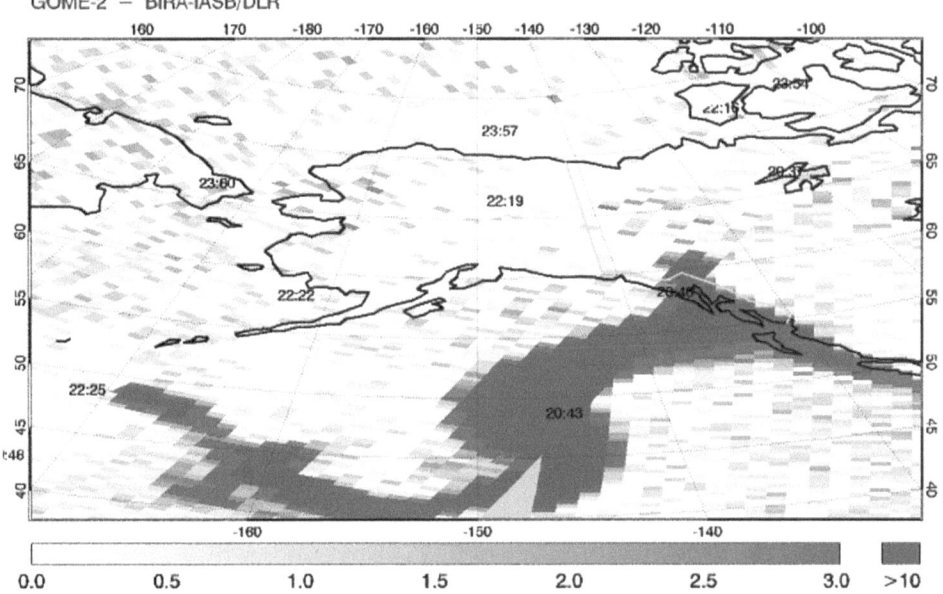

Figure 30. Sulfur dioxide gas concentrations (dark areas) released by the Kasatochi Volcano, which erupted late in the evening of August 7th 2008, as measured 3 days later by Europe's Metop weather satellite. The volcano is in the Andreanov Islands (part of the Aleutians chain) to the left of the gas cloud. Credit: BIRA-IASB/ESA.

volcanic ash cloud. One such image was released by NASA of the Icelandic ash cloud, taken by the experimental Calipso satellite, but that single satellite passes over the volcano too infrequently to provide regular information.

Of the 60 or so volcanic eruptions worldwide each year, only a handful are being monitored on the ground, so this space-based alert service is often the first sign that a problem may arise. Ground and airborne instruments can then be deployed to assess the risks in detail. Nine volcano information centers around the world divide up the globe among themselves and provide alerts when a volcanic eruption occurs in their area, with coordination provided by the center in Washington, DC.[19]

The picture of Antarctica in Figure 31 is made up of over 100 images taken by Landsat and other NASA satellites. The resulting mosaic has then been computer enhanced to highlight the terrain. Before the space age, the interior of Antarctica was

---

[19]   Grimston and Haslam (2010); the nine Volcanic Ash Advisory Centres are in Anchorage, Buenos Aires (Argentina), Darwin (Australia), London (UK), Montreal (Canada), Tokyo, Toulouse (France), Washington, DC (USA) and Wellington (New Zealand) – see *www.ssd.noaa.gov/VAAC/vaac.html* and *http://sacs.aeronomie.be/*; the NASA Lidar image is at *www.nasa.gov/images/content/446297main_apr17-calipso-hi.jpg*.

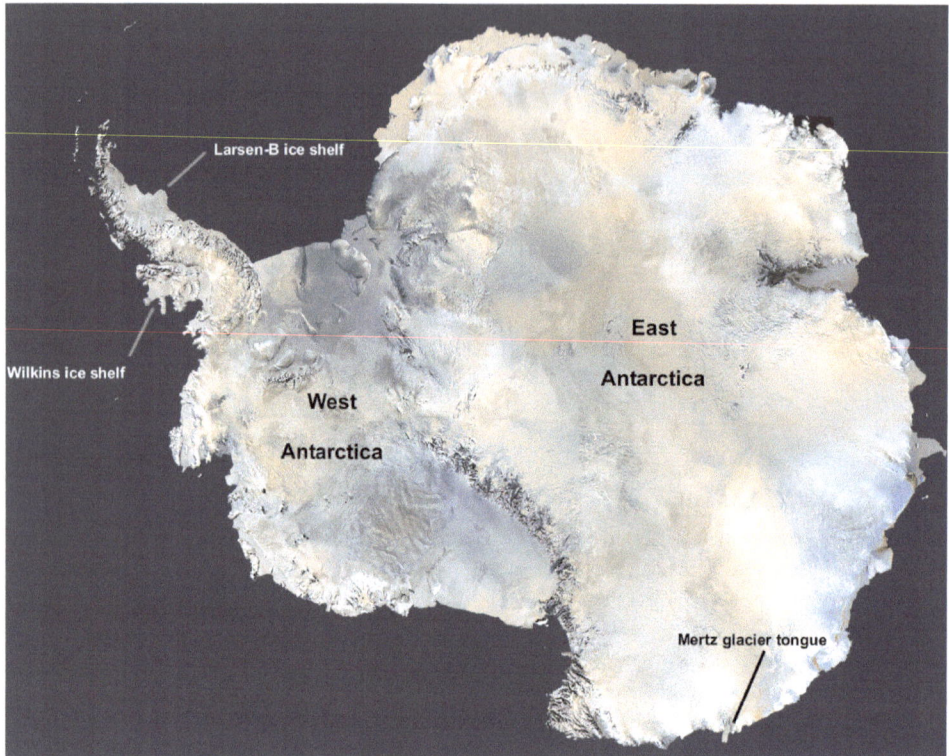

Figure 31. Antarctica. © Fugro NPA Ltd, *www.fugro-npa.com.*

known only in outline, but now satellites monitor it regularly. Comparing images taken years apart highlights changes. Figure 32 is an imaging radar picture taken by the Envisat satellite showing in more detail the ice tongue that can be seen in the lower right of Figure 31. The tongue is the extension of the Mertz glacier into the ocean. A crack across the center of the ice tongue visible in the image eventually gave way in February 2010 when the large iceberg to the right in the image crashed into it. The part of the ice tongue seaward of the crack is about 80 km in length.

Another visual display of melting ice in Antarctica is shown in a dramatic sequence of images taken by NASA's Terra satellite in 2002. Figure 33 shows the Larsen-B ice shelf collapsing into the Weddell Sea – Larsen-B is near the tip of the Antarctic Peninsula that sticks outwards and upwards on the left of Figure 26. The first image shows the floating ice shelf covered in fine blue lines that are ponds of melt-water on the surface. In the second image, 17 days later, a section measuring about 800 sq km has broken off. The third image a week later warned scientists that dramatic events were happening because the blue lines had started to disappear – the melt-water was chiseling its way right through the ice shelf. The last two images 10 and 12 days later show the whole ice shelf shattering into thousands of icebergs. A piece of ice the size of Rhode Island state had disappeared.

Satellite-borne imaging radars have captured several other major ice shelf

Figure 32. The Mertz glacier tongue in December 2007 stretching out into the Southern Ocean on the Adélie Coast of Eastern Antarctica. Two years later, the large whale-shaped iceberg on the right smashed into it and snapped it off at the partial break line (see text). The image is about 300 km from top to bottom. Credit: ESA.

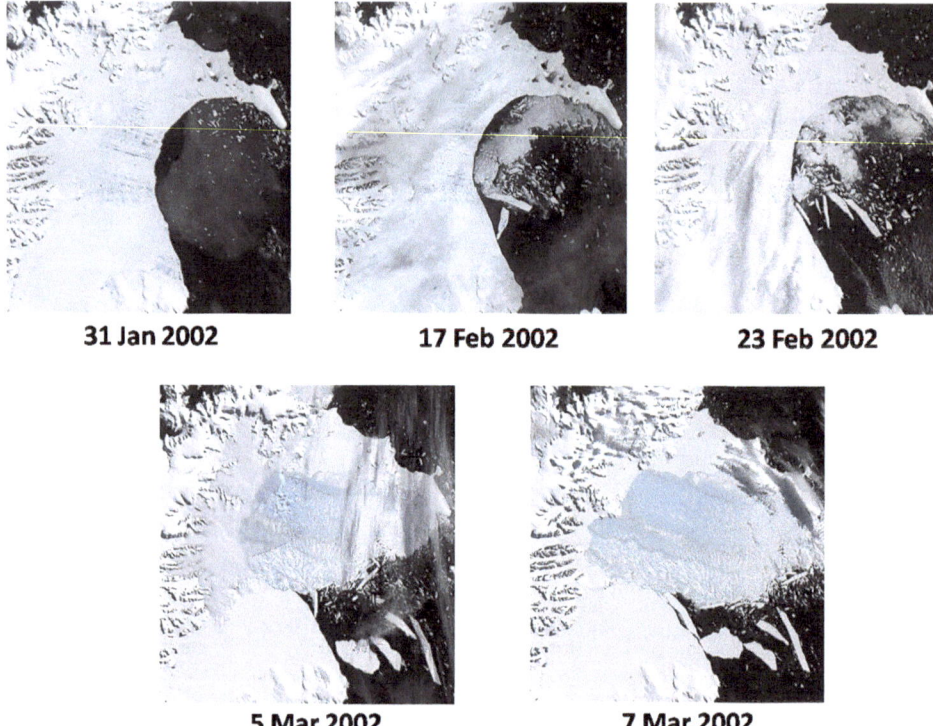

31 Jan 2002   17 Feb 2002   23 Feb 2002

5 Mar 2002   7 Mar 2002

Figure 33. Collapse of the Larsen-B ice shelf in 2002 as captured by NASA's Terra satellite. Credit: NASA/GSFC Scientific Visualization Studio.

disappearances in Antarctica, even in mid-winter, when it is permanent nighttime. One ongoing example is the gradual break-up of the Wilkins ice shelf just to the south of Larsen-B on the Antarctic Peninsula.[20]

When floating ice like Larsen-B or the Mertz tongue melts or breaks off, sea level doesn't rise – just as the level of an iced drink stays the same before and after the ice in it melts.[21] Of course, it suggests that something is warmer than before – either the sea beneath the floating ice or the air above it. Perhaps more importantly, the collapse of Larsen-B caused the land-based glaciers that feed it to speed up in their glacial migration towards the sea – they had been uncorked.

Unlike floating ice shelves, when ice disappears from the land, sea level does rise. One of the more worrying findings in recent years is that Greenland (in the Arctic) and West Antarctica were much less ice-bound in geologically recent times. Samples

---

[20]   Perhaps the best of several image sequences can be found at *www.esa.int/esa-mmg/ mmghome.pl* and search using "Wilkins Ice Shelf" as keyword.

[21]   There is in fact a small increase in sea level due to the melting freshwater being less dense and thus slightly more voluminous than seawater.

drilled out from the base of Greenland's ice sheet show that it was covered in trees 400,000 years ago. The disappearance of Greenland's ice would raise sea levels more than 7 m (24 ft). The evidence in West Antarctica from samples from under the ice is that it, too, was relatively ice-free in about the same time period. West Antarctica is the left-hand side of the continent in Figure 31, leftward of the brown line that snakes from bottom center to top left of center – the part that looks a bit like an elephant's mouth and trunk. Sometimes called Lesser Antarctica, if its ice were to melt, sea levels would rise about 5½ m (19 ft) around the world.

Fortunately, the rest of the continent, Eastern or Greater Antarctica, seems to have been ice-covered for many millions of years and is unlikely to melt, which is just as well, since if it did, sea levels would rise more than 50 m (170 ft), namely 6 m above the torch on the Statue of Liberty. Drilling into the surface and pulling out the "core" makes it possible to analyze the layers of earth beneath Antarctica – the deeper the core beneath the surface, the older the information it contains. Recent results hint at the possibility that even East Antarctica might in the distant past have oscillated between ice-free and ice-covered from time to time, but the evidence is inconclusive.[22]

Satellites are really the only way to measure the extent of the Greenland and Antarctic ice masses using altimetry, imagery of moving features, 3D images and, most recently, gravity measurements. Since altimeters were first flown on satellites in the 1960s, scientists have been monitoring glaciers in both polar regions and in the world's "third polar region", the Himalayas, and other mountain regions.

Another piece of the Antarctica puzzle was fleshed out by satellite altimetry in the 1990s. Using seismic measurements and ground-penetrating radar, scientists suspected the existence of lakes on the solid surface 2 miles below the glaciers in central Antarctica. Europe's ERS-1 satellite mapped the continent with its altimeter and showed the existence of Lake Vostok – a 250-km-long lake the size of Lake Ontario buried at the foot of the ice cap. The flat surface of a lake shows up clearly in an altimeter or an imaging radar because its smooth reflection stands out in comparison with the jagged reflections from dry land. But it is something of a mystery how the smooth echo is captured on the surface of the glacier 4 km above the lake. In any case, Figure 34 shows an imaging radar view of Lake Vostok taken by Canada's Radarsat.

When first discovered, Lake Vostok was thought to contain pristine water untouched for millions of years, but recent research has shown the existence of a network of lakes below the ice – more than 160 have been detected to date, although Lake Vostok is by far the largest. The water in these lakes is slowly exchanged with the overlying ice, but, worryingly, from time to time, there is a relatively rapid transfer of water – in one case, the ice sank by 3 m in one area and rose by 1 m in two areas about 150 km down-slope, indicating the existence of a river between the two

---

[22] Bell (2008), and Naff (2010); the Larsen-B images are online at *http://svs.gsfc.nasa.gov/goto?3123.*

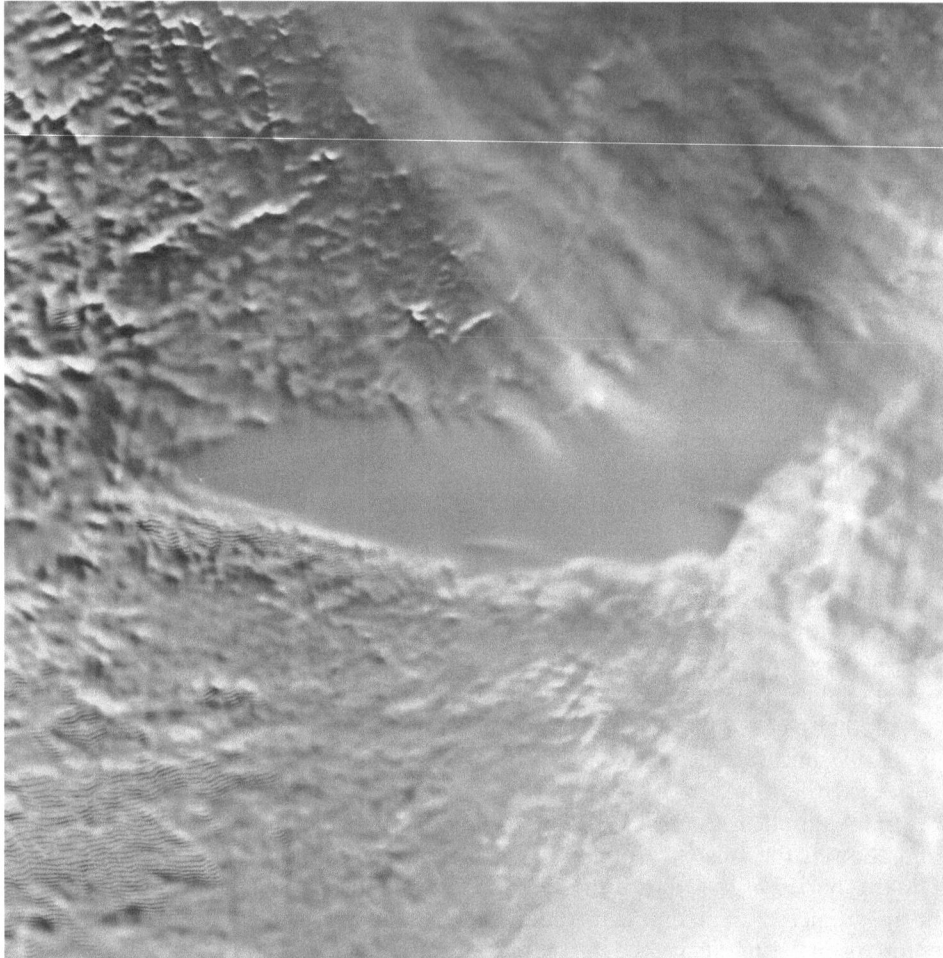

Figure 34. Lake Vostok as viewed by the imaging radar on Canada's Radarsat. Credit: NASA/Goddard Space Flight Center Scientific Visualization Studio. Additional credit goes to Canadian Space Agency, RADARSAT International Inc.

areas. The existence of a plumbing system under the ice is just one more thing to monitor if we are to measure and understand changes in the ice.[23]

On the ground and underwater expeditions are no less important in the space age. They provide the detailed snapshot of a few places that complements and underpins the large-scale picture obtained from space. The first man to walk to the North Pole unaided, Pen Hadow, went back in 2009 with two companions pulling a ground-penetrating radar to measure ice thickness over a 1,000-km track. "The only way to accurately gauge the thickness of the polar ice cap is to physically go out there," says

---

[23] Bell (2008).

Hadow. Data collected by US Navy submarines during the Cold War have been released for use by scientists and provide a wealth of information about polar ice thickness since 1950. These and other occasional human and submarine expeditions across or below the Arctic provide ice thickness information that can be compared with the regular and comprehensive but approximate information from satellites, providing an essential sanity check on the satellite data.[24]

OTHER CHANGES

The UN recognized the need to coordinate the measurement of climate information worldwide. The first step was to identify the things to measure and then the second step was to persuade one or more countries to do so. The first step proved quite tricky because the number of things that could be measured is almost unlimited. Another difficulty is that some of the things you might like to measure are beyond the ability of our current technology. Over the past 15 years, climate experts have gradually built up a list of the things they would like to see measured and that it is feasible to do so. The information can then be fed into computer models of the world's climate to see what lies in store for us.

"Essential climate variables" are what the climate experts call the things they want to measure – a "variable" is something that can vary, like the wind or the temperature, as opposed to a "constant" that doesn't change, like the length of the day or the force of the earth's gravity. There are about 40 of these climate variables and you need satellites to measure about two-thirds of them – the official list of the ones that depend on satellites is in Table 3.

Most of the things listed in Table 3 are deceptively difficult to measure. We have already discussed the difficulty of getting accurate temperature and ice measurements over periods of decades. Another difficulty is hidden within words like "land cover" in Table 3. This is a catch-all phrase that refers to the spread of deserts, the cutting down of forests, the increase in urban sprawl, the intensification of farming and other ways in which our use of land might affect climate. Let's look in a bit more detail at how satellites can help answer just two of those elements of "land cover". Are the world's deserts expanding? How fast are tropical forests disappearing?

The IPCC predicts that deserts will expand in Asia, encroaching further into existing pastoral areas and farmland. Deserts in northern China are spreading at twice the rate of a generation ago.[25] The spread of deserts could be caused by human actions such as over-grazing of fragile grasslands that border the deserts, or it could be due to changes in climate. In either case, it is important to monitor the situation. Watching from space is the cheapest, fastest and most effective way to do this.

Forests help to soak up ("sequester" is the technical term) carbon dioxide, so, for that reason alone, we should be worried by their disappearance. Tropical rainforests

[24] Cooper (2009); NASA website *www.nasa.gov/topics/earth/features/seaice_skinny.html*.
[25] Cruz *et al.* (2007) p. 486, and Mead (2008) p. 325.

Table 3. Essential climate variables largely dependent upon satellite observations.[26]

Domain	Essential climate variables
Atmospheric (over land, sea and ice)	Precipitation, earth radiation budget (including solar irradiance), upper-air temperature, wind speed and direction, water vapor, carbon dioxide, ozone, aerosol properties
Oceanic	Sea-surface temperature, sea level, sea ice, ocean color (for biological activity), sea state, ocean salinity
Terrestrial	Lakes, snow cover, glaciers and ice caps, albedo, land cover (including vegetation type), fraction of absorbed photosynthetically active radiation (fAPAR), leaf area index, biomass, fire disturbance, soil moisture

still exist in South and Central America, Africa and Asia, but the spread of humans is destroying them at an alarming rate. This subject is discussed further in the next chapter.

Climate change is just one of the ways in which environmental change could make the planet uninhabitable for humans. One analysis[27] identifies eight other ways in which we humans are trying to put the earth out of business, including the following:

- Nitrogen and phosphorus cycles: industrial fertilizers pollute water and create hypoxic "dead zones".
- Biodiversity loss: land development is causing one of the great mass extinctions in earth's history.
- Ocean acidification: increased carbon dioxide in the atmosphere makes the sea surface more acidic, weakening ocean ecosystems such as coral reefs.
- Freshwater depletion: rivers are drying up and sub-surface aquifers are being drained.
- Land use: increased conversion of forests to croplands, over-farming that leads to desertification and urban sprawl are some of the worrying trends.

Satellites can help to monitor all of these, but particularly the last two – freshwater and land use. We will return to these two topics in Chapters 4 (land use) and 5 (freshwater).

THE SUN

This book is about the many ways in which the earth is being watched from space. In this chapter on climate change (and nowhere else), two roles that satellites play by

---

[26]   Mason and Bojinski (2006) p. 8.
[27]   Foley (2010).

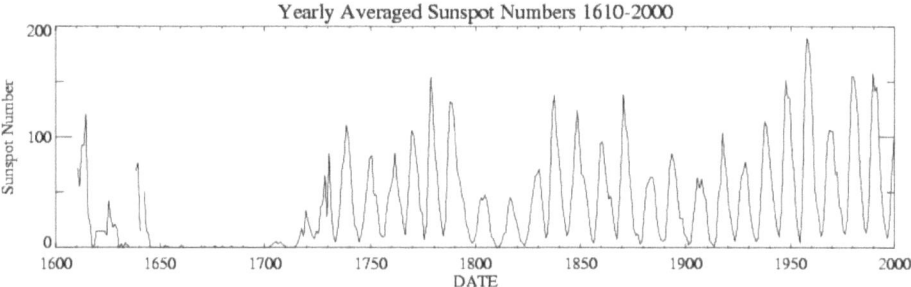

Figure 35. The number of sunspots outlines the 11-year solar cycle. The 70-year gap in the 17th century is evident on the left of the diagram. Credit: NASA.

looking upwards and outwards from the earth will be described. The first of these concerns our sun because it provides the heat and light that support life on earth. A small increase in the energy from the sun would produce the global warming described above and a small decrease would cause an ice age. So an important part of understanding climate change is to check the amount of radiation from the sun.

Measuring the energy from the sun precisely is quite difficult from the ground. The atmosphere absorbs or reflects a lot of the sun's energy, such as when clouds prevent direct sunlight reaching the ground. Changes in the diameter of the sun and in the number of sunspots (dark spots on the sun's face) and faculae (bright spots) have been suggested as indicators of the sun's energy. It has proved impossible to reliably detect changes in the sun's diameter – some astronomers claim to have measured significant changes but other astronomers have been unable to duplicate these results. Recent measurements of the sun's diameter from space have shown the changes to be so small that they would be undetectable on the ground and seem not to be useful for checking the sun's energy output.[28]

Sunspots vary in number dramatically over an approximately 11-year period (see Figure 35). Sunspots are caused by the sun's magnetic field exposing slightly cooler and thus darker material than the rest of the sun's surface. So, if there are lots of sunspots, it suggests that the sun's magnetic field is particularly active and perhaps that results in more energy reaching the earth. Likewise, the absence of sunspots suggests that the sun's magnetic field is relatively quiet and perhaps that means that earth is receiving less energy. Sunspots seem to have been absent from the sun for a 70-year period from AD 1645–1715 (see Figure 35) and this falls within a period known as "the Little Ice Age". However, the Little Ice Age seems to have been confined to Europe and lasted from about AD 1300 to 1850, which is much longer than the period without sunspots, so it is something of a leap of faith to say that reduced sunspots means lower temperatures. Nevertheless, the supporters of this theory claim to see a general, if irregular, increase in each 11-year cycle of the

---

[28]   Marcelo *et al.* (2010).

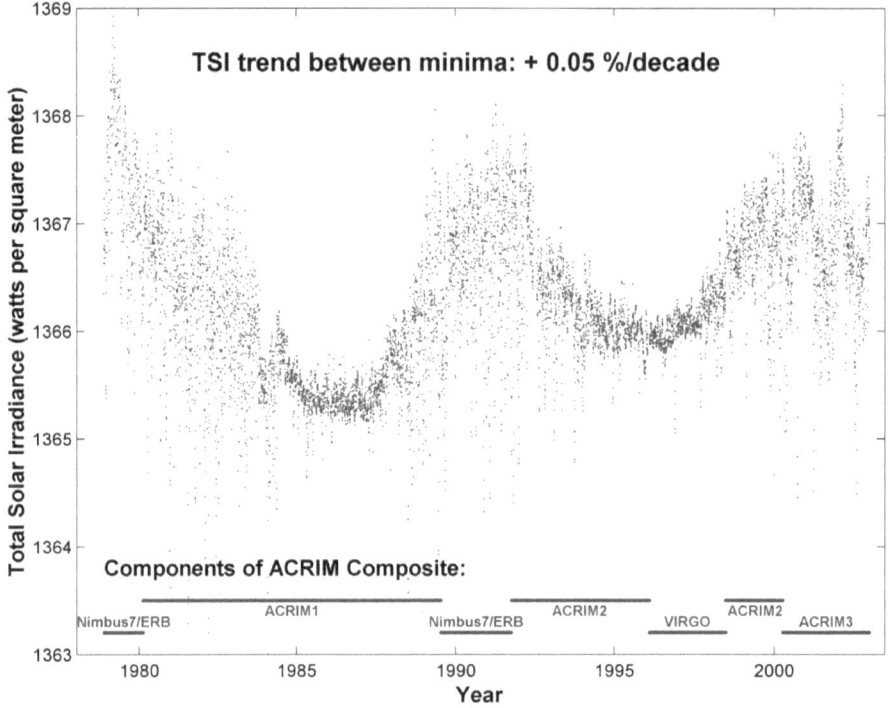

Figure 36. Measurements of solar energy reaching the earth by seven satellites since 1978 – the period of each satellite's data is indicated by the bar above the date. Credit: NASA, ACRIM3 Science Team (R. Willson).

sunspot activity since about 1850 and find this curiously coincident with the increase in global warming. Others have noticed a slight reduction in the length of the 11-year cycle from about 11.5 years in the 1700s to about 10 years in the late 20th century. They consider that a shorter cycle means that the sun is more active and intense. Recent results suggest that the small change (about a quarter of 1%) in the sun's energy over the 11-year cycle is enough to influence the winds in the Arctic region and thus create cold weather for northern latitudes.[29]

The "sunspots cause climate change" enthusiasts are a long way from proving their case, but that shouldn't stop us looking for changes in the sun's energy. Special instruments have been flown on satellites for the past 30 years to do just this. It has been difficult to get instruments on two different satellites to give exactly the same reading at the same time so it is too early to make confident statements about the sun's energy. Figure 36 shows an amalgamation of data from seven satellites, starting with the Nimbus7 weather satellite in 1978 and ending with

[29]   Friis-Christensen and Lassen (1991), Shindell *et al.* (1999), Rapp (2008) pp. 143–209, and Clark (2010).

the ACRIM3[30] instrument on the ACRIMSAT satellite dedicated to measuring the sun's energy from 2000 onwards. The compilers of these data say that they show a small increase in solar energy – one-twentieth of 1% per decade. However, the IPCC says that there is "no significant long term trend".[31]

One of the big difficulties in this affair is that the energy we receive from the sun varies during the year by about 7% just because the earth's orbit is not quite a circle. Earth is about 152 million km from the sun in July and only 147 million km away in January – meaning that the southern-hemisphere summer is a little more intense than that in the northern hemisphere, and its winter, too. The sun's energy seems to vary by a tenth of a percent or so during the 11-year sunspot cycle, as shown in Figure 36.

So, the instrument on a satellite has to measure the 7% change over the course of the year and a 0.1% change over the 11-year cycle and only then try to detect a long-term change that might affect earth's climate. There haven't been enough 11-year cycles since the satellite measurements began for a clear upward or downward trend to emerge. So it is likely to be a few more decades before we get conclusive information about changes in the sun's energy.

THE ASTEROIDS

Climate change made the dinosaurs extinct about 65 million years ago – along with three-quarters of all life forms! Dinosaurs had inhabited the earth for about 150 million years so it must have been one heck of a change in the climate that drove them to extinction.

It seems that a 10–15-km-wide asteroid or comet hit the earth at a speed of 20 km/s near the Yucatan Peninsula and caused widespread fires, tsunamis and earthquakes followed by years or even decades during which smoke and ash blocked the sun's rays, giving rise to a period of "global winter". The idea of asteroid impacts was deemed crazy when Velikovsky was writing about it in the 1950s, but the more thorough investigations of Nobel Prizewinning physicist Luis Alvarez and his geologist son Walter in the 1980s made the theory respectable. An alternative theory involving greenhouse gases released by the volcanic processes that created the Deccan Traps in India seems now to be discounted – the vast Deccan lava flows emerged relatively slowly and over long time periods so that the earth could adjust to the effects.[32] In North America 12,900 years ago, a comet or asteroid seems to have thrown up huge amounts of debris and caused widespread cooling. This event may have contributed to the extinction of the mammoth and other animals, and would have been witnessed by humans.[33] More than 1,000 comets have been found that travel right through the solar system to pass close by the sun – they are called sun-

---

[30]   ACRIM = Active Cavity Radiometer Irradiance Monitor.
[31]   NASA (2003), and Solomon *et al.* (2007) p. 30.
[32]   Schulte *et al.* (2010).
[33]   Kennett *et al.* (2009).

Figure 37. Meteor Crater, just south of Interstate 40, a few miles from Winslow, Arizona. Credit: Science @ NASA.

grazing comets – so the possibility of one bumping into the earth by chance cannot be dismissed.

The asteroid threat is not just something out of pre-history. About 50,000 years ago, a 40-m-diameter object smashed into the ground to create the 1.2-km (¼-mile)-wide Meteor Crater (see Figure 37) in Arizona. Figure 38 shows a small piece of the 200,000 hectares (800 sq miles or ¼ million acres) of forest leveled in a radial pattern by a body (probably a small comet traveling at 50,000 km/h) about 40 m wide exploding over Tunguska in Siberia in 1908 – similar in scale to a nuclear bomb blast hundreds of times more destructive than the Hiroshima atomic bomb. In 1994, we saw a comet smash into Jupiter, blasting holes in its atmosphere the size of the earth. The Hubble Space Telescope recently snapped a picture of the aftermath of two asteroids colliding in outer space. Celestial collisions do happen.

The 1908 Siberian blast at Tunguska caused no casualties as far as we know, but if it had fallen in the ocean, a massive tsunami would have resulted – and the ocean makes up two-thirds of the earth's surface, so odds are that's where the next one will land. If it had occurred over New York City, the entire Metropolitan area would have been razed. Collisions of celestial bodies seem to be relatively common and an impact similar to Tunguska is predicted to happen every few hundred years.

It might not destroy Manhattan but the relatively small meteor that landed in Peru in September 2007 left a crater 13 m wide and 15 m deep. It landed in a deserted field but if it had hit the nearby village, it would certainly have demolished a house

Figure 38. Tunguska trees flattened like matchsticks. Credit: The Leonid Kulik Expedition.

or two. The meteor is thought to have been 1–2 m in size, traveling at about 7 km/s. Objects of this size strike the earth every year, on average. Smaller meteorites hit several times a year – for example, a fist-sized rock slammed into a Doctor's surgery in Virginia in January 2010, and the websphere is replete with videos and images of a meteorite streaking across the mid-West before exploding over Wisconsin 3 months later.

NASA has been tasked by the US Congress to identify 90% of asteroids and comets in the inner solar system bigger than 140 m by 2020. A special camera is being installed on a mountain top in Hawaii to scan the sky for these faint objects. The European Space Agency's Gaia satellite will help complete the survey when launched in 2012. As a side effect of its main mission to map a billion stars in our galaxy, Gaia will identify and locate thousands of asteroids and comets. Given that the Tunguska and Meteor Crater events were caused by objects just 40 m wide, you might wonder why Congress has only instructed NASA to seek objects bigger than 140 m. The 140-m figure seems to have been selected as being affordable and, of course, NASA will take note of any smaller objects detected, but that won't be anywhere close to 90% of them. Until humans are killed, or major damage caused, by an object from outer space, it seems likely that detection of the smaller bodies will remain a low priority for the world's astronomers.

Hollywood's answer to an approaching asteroid is to blast it with hydrogen bombs. The more prosaic and effective response is to hover a space probe close to the asteroid way out in space and let the probe's gravity pull the asteroid gradually off its earth-bound path. This process might take anywhere from a year to a decade, depending on the size of the asteroid and how near to full-on the impact would otherwise be, so early warning is essential. If we only get a few weeks' warning, a hydrogen bomb may be the only hope, even though its effectiveness against the rubble-like consistency of many asteroids and comets is doubtful.[34]

---

[34]   National Research Council (2010), and Stone (2008).

4

Commercial surveillance: mapping on a large scale

INTRODUCTION: FROM LANDSAT TO GEOEYE

There are two distinct ways in which the commercial world is involved in observing the earth from space. One way is for a commercial company to build an earth-observing satellite and sell the images to anyone willing to pay for them – whether they are in the public or the private sector. Companies such as GeoEye, Digital Globe and SPOT Image fall into this category.

The other form of commercial involvement is for a private-sector organization or individual to buy space imagery that may come from public or commercial satellites. The market has evolved to the point that these images are used by a wide variety of organizations and people to help with an almost unlimited range of problems. It would be impossible to describe all the uses people make of the internet or the cell phone, and the uses of satellite images are also so wide and ever widening that a comprehensive review would be impossible. In this chapter, we will look at probably the most mature commercial use of satellite images, namely farming and forestry, to illustrate the potential of the data. The range of other uses will be indicated by shorter looks, including 3D maps, oil spills, mineral exploration and construction.

First, we will discuss how private companies have taken on a role previously monopolized by the military.

Since 1972, relatively detailed imagery from satellites has been available for the general public to purchase. Before that time, weather satellite images were freely available but showed details that were on such a large scale that only the weather forecasters were interested. The military in the USA, the Soviet Union and their allies had been using detailed satellite imagery for a decade by that time, but the images were so sensitive that they were shown to only a small community of intelligence analysts. Astronauts had taken spectacular pictures as they orbited above the earth but these images were infrequent and random in time and place.

P. Norris, *Watching Earth from Space,* Springer Praxis Books,
DOI 10.1007/978-1-4419-6938-5_4, © Springer Science+Business Media, LLC 2010

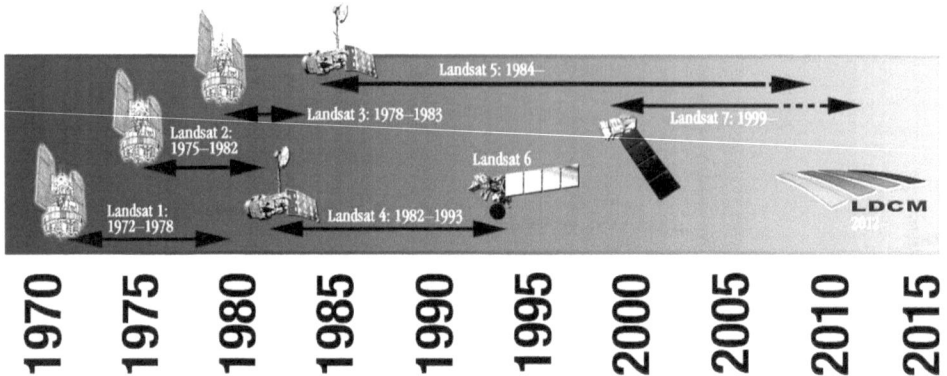

Figure 39. NASA's Landsat series: the first non-military surveillance satellites. Credit: NASA.

NASA launched the first of what were later called the Landsat satellites[1] in 1972. It was widely recognized as a ground-breaking initiative and that assessment has stood the test of time. At first glance, the quality of its images seemed slightly disappointing, with resolution (pixel size) of about 60 m (200 ft). However, the images came in four colors or "bands", and this allowed analysts to distinguish different types of crops, to spot polluted waters and so on. As the NASA website puts it, "Landsat sensors have a moderate spatial-resolution. You cannot see individual houses on a Landsat image, but you can see large man-made objects such as highways. This is an important spatial resolution because it is coarse enough for global coverage, yet detailed enough to characterize human-scale processes such as urban growth".

If you were happy with field-sized details, then the images showed the earth in ways that hadn't been seen before. A total of six Landsats have now been placed in orbit[2] (see Figure 39) and the later versions have become gradually more sophisticated so that the latest, Landsat-7, gives images with 15-m resolution in black and white or 30-m in color.

There are now more than 100 satellites in orbit, sending back images similar to those that the first Landsat took. Some show much more detail than Landsat, but almost all copy its basic characteristics. The images are taken using an electronic, digital camera – not a wet film that some military spy satellites still use, and thus not limited in time by how much film is onboard. The images are stored in the satellite until, perhaps a few hours later, a friendly ground station comes into view, whereupon they are transmitted by radio link to the ground – the wet-film approach requires the use of a returning capsule.

Landsat covers the whole of the earth, except for the extreme north and south

---

[1]   The first satellite was originally called the Earth Resources Technology Satellite (ERTS).
[2]   Landsat-6 was lost when its launcher blew up.

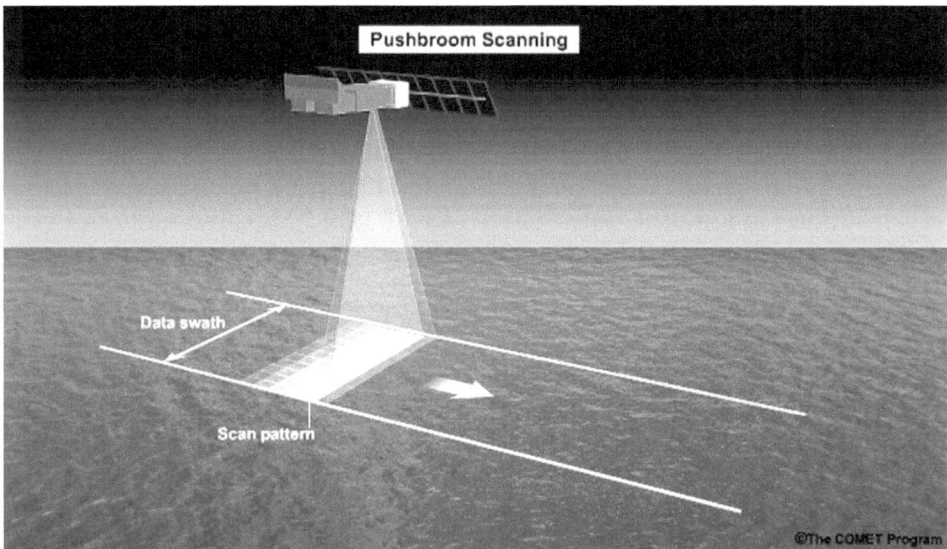

Figure 40. The satellite, moving from left to right, sweeps out with its pushbroom camera an image of a strip or swath of the earth below. Credit: UCAR.[3]

polar areas. The satellites are in an orbit that is carefully chosen to pass over an area at the same time of day. It may take 2 or 3 weeks to come over again, but when it does, it will be at the same hour. The thinking behind this feature is that the sun will be in the same position in the sky each time, so that shadows and glare will be consistent. If you take images at other times of day, the difference in glare and shadowing might show changes that are just a trick of the light. This "sun-synchronous" feature, as it is called, requires the satellite to be between 500 and 1,000-km altitude.

A peculiar trait of almost all such satellites is that they don't take a picture in the sense that your personal digital camera does. Instead, they sweep out a continuous strip as they fly over the ground. Your digital camera has an electronic device called a CCD[4] that takes a rectangular picture when you press the shutter. The electronic device on the satellite has a long thin CCD or some similar device that takes only a line or two of a picture when its shutter is pressed. The satellite moves inexorably forward over the ground, so a thousandth of a second (a millisecond) later, it is 8 m further on and it opens the shutter again to snap another line or two. By choosing the number of lines in each snap and the exact time between snaps, the satellite

---

[3]   The source of this material is the COMET® website at *http://meted.ucar.edu/* of the University Corporation for Atmospheric Research (UCAR), sponsored in part through cooperative agreement(s) with the National Oceanic and Atmospheric Administration (NOAA), US Department of Commerce (DOC). © 1997–2010 University Corporation for Atmospheric Research. All Rights Reserved.

[4]   CCD = Charged Couple Device.

gradually builds up an image of the ground it has passed over. NASA's website has an animated graphic that shows this in action[5] and Figure 40 illustrates the principle. The name "pushbroom" comes from the way the satellite "sweeps" out the image.

The pushbroom technique provides perfectly good images for almost every application. However, an "image" is an artificial idea in the pushbroom world – theoretically, an "image" could go on and on indefinitely. In practice, the images are usually defined as a roughly square strip – length of strip equal to width of swath.

The USA had a monopoly on the commercial provision of satellite images until 1986, when the first of the French SPOT satellite series was launched. The French satellites have generally been more reliable than the Landsats, many of which have suffered partial or total failures. Perhaps the Byzantine management arrangements for Landsat are to blame – responsibility and funding are split between NASA, NOAA, the US Geological Service and (for a while) a private company. The NASA website admits that by the time of Landsat-5, the management was a "shambles" and its "standards languished".[6] A similar "shambles" occurred in the weather satellite arena, in which, again, both NASA and NOAA shared responsibility and funding (see Chapter 2), so the problem seems to be pervasive. The successor to Landsat-7 involves a mix of NASA and the US Geological Service, which sounds as if the earlier lessons about the perils of splitting responsibility may not have been fully learned. It is currently called the Landsat Data Continuity Mission (LDCM), although I suspect that it will become known as Landsat-8 after its launch, which is currently promised for December 2012.

France stuck to a relatively simple management structure when developing its SPOT satellite, although it could have been much more complicated had Germany not refused to help finance it. In 1975, European countries set up the European Space Agency and agreed that new space projects would be offered to that Agency rather than being developed by a specific country. Only if the other countries that financed the Agency declined the offer would a country develop it alone. In 1977, SPOT was offered by France to the Agency alongside an imaging radar satellite called ERS. After much debate and negotiation between the 10 countries, ERS was accepted as an Agency project but SPOT was declined, leaving it to France to develop on its own or not at all. The result was that the first SPOT satellite was launched in 1986, while the first ERS took 5 years longer.

Each of the countries in the European Space Agency is a bit like an owner in a condominium. Each can influence the decisions of the group but may be outvoted. Unlike a condominium, if a country doesn't wish to fund a particular project, it doesn't have to – the other countries fund it and the non-funder has to pay later if it wants to use that satellite or rocket. The high cost of space projects leads inevitably to debate among the countries as to the exact purpose of the project, what it should cost, which companies will undertake the work (each country has its own favored industry) and so on. So, debate begins with the countries deciding whether they will

---

[5]   http://earthobservatory.nasa.gov/Features/EO1/eo1_2.php.
[6]   http://landsat.gsfc.nasa.gov/about/landsat5.html.

join a project. To avoid lengthy deadlock, a project usually begins provisionally, in which perhaps 5–10% of the funding is provided by the countries – just enough for the cost of the full project to be worked out in detail. Then comes the hard debate about putting up the large sums needed to build and launch the satellite.

It's good to talk – but the more parties in the discussion, the longer it takes to reach agreement. Landsat in the USA and ERS in Europe suffered from this problem of "too many cooks spoil the broth", which can continue even after an initial agreement is reached. If the project encounters problems, the debate begins again – delete some part of the project, put in more money, change the contractor, bring in another partner to pay some of the cost, change the management and so on.

The decision not to make SPOT a European project was influenced by the strong opposition of Germany to it. The German Government Minister, Hans Hilger-Haunschild, was nervous about over-loading the newly created European Space Agency, which had already been given the job of developing telecommunications, weather and scientific satellites and a laboratory to fit in the US Space Shuttle, and was now being given the ERS imaging radar satellite. He felt confident that France would not go it alone on SPOT, and that therefore it would be put on hold and could be reviewed in a few years' time. He had also recently agreed to create a second European agency called Eumetsat that would own and operate the weather satellites that the European Space Agency was developing. Germany would be the largest funder of Eumetsat, so Haunschild felt that he had done enough *Europeanizing* of space for the moment. Several countries including Germany also felt that SPOT was a quasi-commercial project that would be inappropriate for the research-oriented European Space Agency to undertake – the ERS radar project, on the other hand, was seen by these countries as a one-off. ERS's imaging radar was also seen by some countries such as the UK as having more military benefits than SPOT.

The French space agency CNES[7] was nervous about its future, having been criticized for over-spending and under-performing. It was therefore motivated to take charge of a significant venture such as SPOT that could become its flagship project and argued strongly inside France for that. CNES ensured some degree of international participation in SPOT by negotiating a small amount of funding from Belgium and Sweden.[8]

In the event, SPOT went ahead as a French project and has continued to the present day; the 1-ton SPOT-1 was launched in 1986, with its successors growing in sophistication up to the latest, the 3-ton SPOT-5, launched in 2002. The 2½-ton ERS-1 satellite was launched in 1991, its nearly identical successor ERS-2 in 1995 and its more ambitious 8¼-ton Envisat successor in 2002. Thus, ERS was far from being a one-off and, indeed, two satellites of an operational imaging radar system called Sentinel-1 are now under construction, financed by the EU and ESA. Images from ERS had a resolution of 26 m and from Envisat of about 30 m. The Sentinel-1 satellites will provide 5-m-resolution images, making them close to the

---

[7] CNES = Centre National d'Etudes Spatiales.
[8] Roy Gibson, personal communication, plus the author's personal recollections.

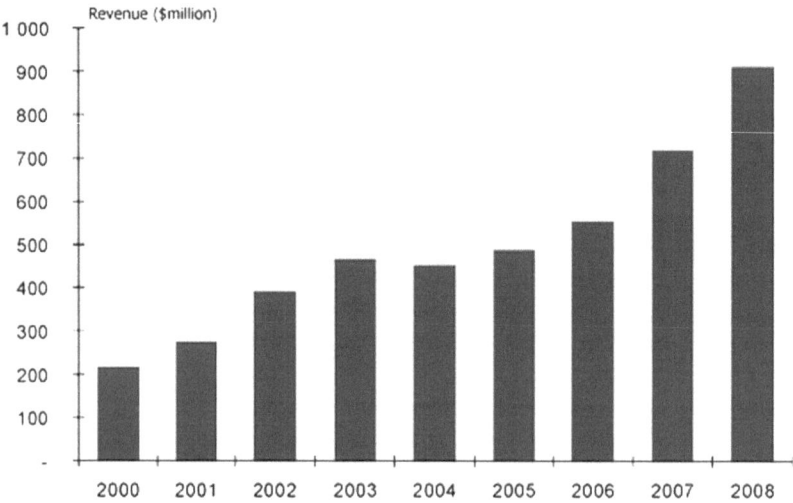

Source: Satellite-based Earth Observation, Market Prospects to 2018

Figure 41. Strong growth of satellite image sales. Credit: Euroconsult.

quality of military and civil–military satellites (Chapter 8). The German and Italian military and civil high-resolution imaging radar satellites to be discussed in Chapter 8 have benefited from the technology heritage of ERS and Envisat.

SPOT is now a joint venture between the French Government and industry, with investment in future satellites increasingly coming from the private sector. France has exploited its investment in SPOT to develop the Helios range of military imaging satellites (Chapter 8) as well as selling commercial satellites to countries such as Chile, Kazakhstan, Taiwan and Thailand.

Since 2008, Landsat images have been available to download free of charge from the US Geological Service website. Previously, they cost $600 a scene, which works out at $3.3 per square km. The demand for the images increased 50-fold within months of the price being removed.[9]

With or without free Landsat data, the world market for satellite images has been growing strongly (see Figure 41). From the moment the first Landsat was launched, enthusiasts have predicted that this market would become huge. Instead, for nearly 30 years, the market sputtered along at around the $100 million-per-year mark. The only truly routine uses of satellite images were by the weather forecasters, who gave their images away free, and the military, who wouldn't give their images away for any price.

The background to the strong growth in the first decade of the 21st century is linked to the removal of military restrictions and is described in Chapter 8. It led to the US Government's decision in 1994 to legalize satellite imagery with pixel size of

---

[9]   Keith (2010b).

half a meter, hitherto the preserve of the military. Russia's decision to sell slightly blurred versions of its spy satellite imagery from 1990 onwards was one of the triggers for the US decision. With the break-up of the Soviet Union, Russia's space program was starved of funds, and the sale of imagery was one of the few ways to attract revenue. The images were all on traditional film and were scanned into computers for commercial sale, just as you might scan in old snapshots to your computer.

The US Ikonos satellite, launched in 1999, was the first to benefit from the US Government's change and was followed soon after by another US satellite called Quickbird in 2001. The two US companies concerned, GeoEye and DigitalGlobe, have turned the market full circle by winning contracts from the US military to supply huge amounts of imagery. To some extent, therefore, the apparent growth in the market is the US military buying data from the market instead of using its own satellites. GeoEye, for example, says that two-thirds of its revenues come from the US Government and that state of affairs is expected to continue.[10] Furthermore, the US Government paid nearly half the $500 million cost of building and launching that company's latest satellite, GeoEye-1, and committed to buying $150 million per year of images as long as the satellite performs well. Digital Globe has a similar arrangement. The role of US military and intelligence agencies in this field is discussed further in Chapter 8.

The images on the front cover of this book were taken by the GeoEye-1 satellite and illustrate the level of detail that this type of satellite provides. On the lower right of the cover, a ship can be seen in Gulfport, Louisiana, washed ashore by Hurricane Katrina. Next to it, forest fires in California are easily located. These two images indicate the power of these detailed images to help emergency workers. The two upper images are more iconic – Ayers Rock/Uluru in central Australia and the Golden Gate Bridge in San Francisco – but they show that ships and cars are readily spotted.

The two US companies no longer have this part of the market to themselves. France's SPOT-5 satellite provides images with 2½-m resolution and its successor, Pléiades, will offer 50-cm imagery when launched in late 2010. Imagery from the Cartosat-2 satellites owned by the Indian Government has 80-cm resolution and Israel's Eros-B1 70-cm or better. Turkey has recently purchased a $350 million 70-cm-class satellite to be launched in 2011.[11] A batch of systems already in orbit offering imagery with 5-m or better resolution include the joint Chinese–Brazilian CBERS-2B, Germany's five RapidEye satellites, South Korea's Kompsat-2 (1-m resolution in black and white), Taiwan's Formosat-2, Malaysia's Razaksat, Thailand's Theos and the list goes on. New entrants expected in 2010 include South Africa's 80-kg ZA-002, Belarus's 400-kg Belka-2 (both of which will offer 2-m-resolution images) and Chile's French-built 150-kg SSOT satellite offering 1½-m-resolution images in black and white, and 6-m in color. But, as already pointed out

---

[10] de Selding (2009b).
[11] Nativi and Taverna (2009).

in Chapter 1, resolution isn't the only thing that matters. In total, 44 earth-observing satellites were launched in the past 10 years, excluding military and weather satellites. A further 180 surveillance satellites are expected to be launched by 2018 according to one industry analyst.[12]

In Chapter 8 on military imaging satellites, we will pick up the story of commercial satellites and see them in the context of their military counterparts. For example, in Russia, India and several other countries, a commercial industry has emerged but the influence of the military is evident. In that chapter, we will also take a look at the recent batch of commercial imaging radar satellites, some of which offer 1-m resolution and which, by their nature, can take images at night and through clouds – both of which are show-stoppers for the satellites mentioned in the previous paragraph.

As mentioned above, the uses of satellite images are so wide that a comprehensive review would be impossible. A selection of commercial uses of satellite images will be described in this chapter in order to indicate the power of the technology. We will start with 3D maps.

3D MAPS

An atlas or map probably gives some idea of the elevation of the ground by using colors to show mountains or by including contour lines and the occasional height of a mountain top. In the last 30 years, this information has proved unsatisfactory in many places in addressing questions such as:

- Where should cell phone towers be sited so as to minimize the number needed and leave no gaps in cell phone coverage?
- Where will flood waters or heavy rainfall flow?

Even in rich Western nations, the accuracy of the existing maps was unequal to these tasks.

The cell phone boom started in the 1980s and soon gave rise to a new industry that produced the maps needed for deciding where to place the many thousands of relay towers. Satellite images were the solution in many countries, allowing elevation information to be calculated across a whole region or country rapidly and automatically. The challenge for the phone companies was to calculate the coverage (and the gaps in coverage) of a network of relay towers. They needed to know the hills and valleys in great detail plus the buildings and trees that might obstruct the cell phone signals. A network of towers to cover the USA could require tens of thousands of installations, each costing between $15,000 and $150,000, depending on the existing facilities. So, reducing the number of towers by a few percent based on careful analysis of coverage could save millions of dollars – 100 towers fewer at an average cost of $50,000 saves $5 million.

---

[12]   Northern Sky Research (2009), and de Selding (2008).

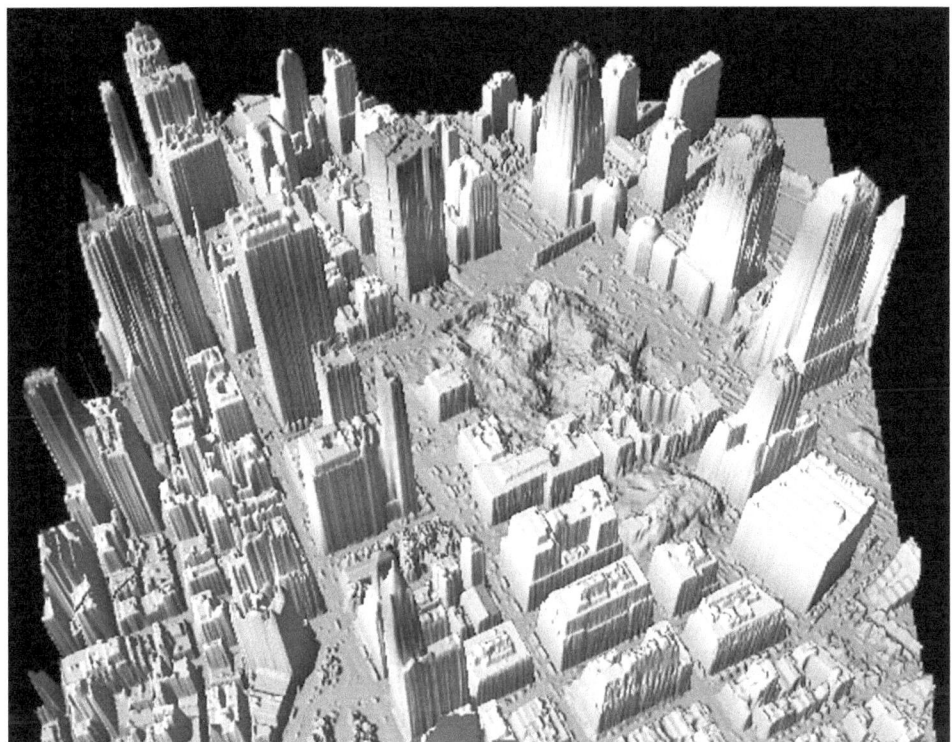

Figure 42. Airborne Lidar image of World Trade Center "ground zero", New York, 12 days after 9/11. Credit: NOAA/US Army JPSD.

Satellite imagery provides an answer especially in rural and open areas. Two images of a scene taken from different perspectives allow the altitude of each point to be calculated – using parallax, just as the human eye does to gauge depth and distance.

Two satellite images of the same area taken from different directions can be computer processed to work out the elevation of each point in the scene.[13] A reference point is located in each image and the difference in location is a measure of the elevation of the point. This process is repeated throughout the scene to build up a map of the elevation contours in the scene. The processing is complex and somewhat approximate, but the result is adequate for many purposes. Elevation accuracy using SPOT images is quoted as better than 5 m, while DigitalGlobe's WorldView-2 satellite is said to provide elevation information that is accurate to 30 cm.[14] The same images can also provide information on the tree cover and other conditions that affect radio waves (see the section on "farming" below), which allows cell phone coverage to be calculated more accurately.

---

[13] Williams (2001) pp. 114–138.
[14] Longhorn (2010c).

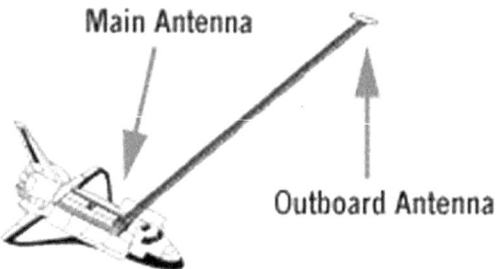

Main Antenna

Outboard Antenna

Reflected radar signals collected at two antennas,
providing two sets of radar signals separated by
a distance.

Figure 43. Shuttle flight 99 in February 2000 carried two radar antennas 60 m apart enabling elevation contours on the ground to be measured worldwide. Courtesy NASA/JPL-Caltech.

In urban areas, even the best satellite images may fail to give adequate maps because tall buildings block the scene behind. Survey companies use aircraft or helicopters fitted with a laser radar, called a Lidar (a word that is loosely derived from light detection and ranging), to accurately map a complex urban area in three dimensions – Figure 42 illustrates the detail this technology can provide. Cell phone coverage in a city is affected not by trees, but by blockage by, and reflections off, buildings. Specialist software has been developed to work out cell phone reception quality using the 3D maps that Lidar and other more traditional techniques provide.

For nationwide coverage, satellites provide the images of choice in most countries. With the arrival of imaging radar satellites, bad weather is no longer a hindrance to producing elevation maps. In fact, an imaging radar that flew on the Space Shuttle in 2000 allowed NASA to produce elevation contour maps for most of the world's land areas in just a single 10-day mission. The resulting maps have a granularity of about 30 m and the accuracy of the elevation contours is about 5–10 m.[15]

The apparent ease with which just 90 h of imaging radar time on the Shuttle could produce an elevation map of about 80% of the world illustrates some of the problems with the technology. The Shuttle set-up comprised two imaging radars separated by 60 m (using a long mast – see Figure 43) whereas other imaging radar satellites carry only one radar. With only a single radar, a satellite has to revisit the scene to take a second image days or weeks or even years later. In between the two images, the ground below may change due to the seasons, the moisture levels of the ground may differ, the cloud cover may vary and the knowledge of the orbit of the satellite may not be known perfectly. Let's look at some of these problems in more detail.

Radar waves can penetrate into frozen snow or ice or very dry soil – this is the basis for ground-penetration radars used by forensic teams to search for buried

---

[15] Farr *et al.* (2007).

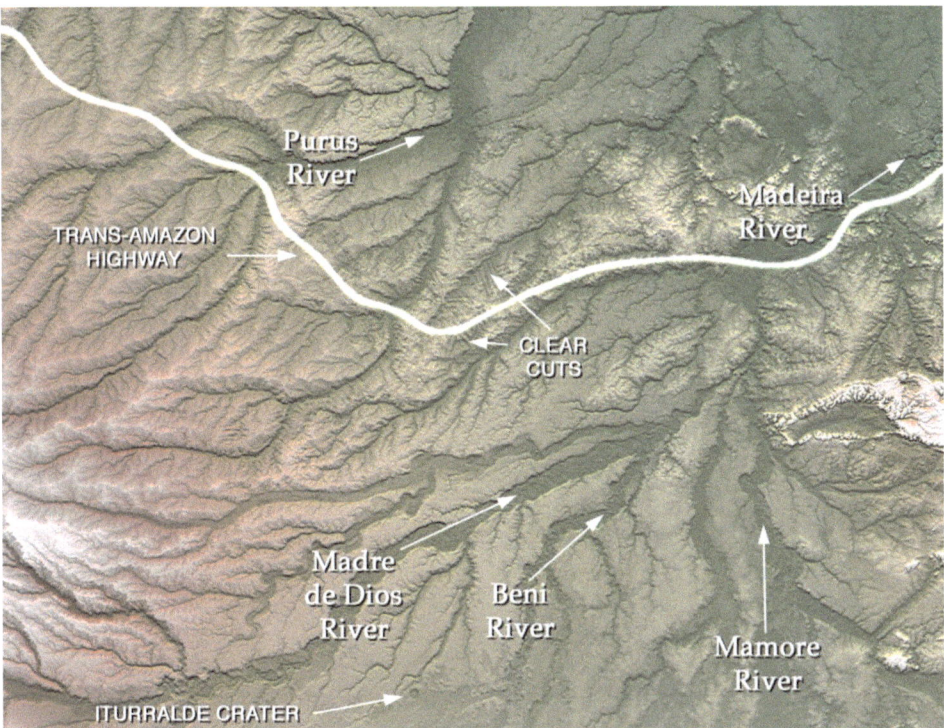

Figure 44. Pando Province, Bolivia, and adjacent parts of Brazil (at the top and to the right) and Peru (to the left), shaded and colored to show the elevation measured by the Shuttle imaging radar in February 2000; the area of the map is about 700 km east–west by 500 km north–south. Courtesy NASA/JPL-Caltech.

bodies or man-made objects or cavities, etc. So, if the snow or ice or moisture level of the ground varies between the two images, an apparent change in elevation will be seen. This is a weakness of the imaging radar approach in that we don't know whether the measured elevation is from the top of the snow pack or the buried ground surface below – much of the northerly terrain was snow or ice-covered when the Shuttle flew in February 2000. A similar uncertainty applies to data from the ultra-dry Sahara Desert, parts of which are known to allow some radar signals to penetrate.

In areas in which the ground is covered with vegetation, the surface from which the radar signal bounces can vary, depending on the height, structure, density and moistness of the plants or trees. If the vegetation is dense enough, as in a tropical forest, the signal is reflected from somewhere in the canopy and not from the ground. As a result, clear cuts in dense forests or jungles are readily noticed (see Figure 44). As a consequence, if the density or moistness of the foliage changes between the two images, then the resulting elevation measurement will be in error.

Heights measured in cities represent average building sizes, rather than the height of the ground on which the buildings sit. Therefore, major changes in building patterns will change the measured elevation.

Although radar can see through clouds and rain, the signal is affected to some extent and in particular is slowed and bent slightly. The slowing of the signal in cloud or rain means that an object appears further away than if the air were dry. The bending of the signal means that the object is slightly offset from its position if the sky had been clear. In both cases, the effects are usually of the order of a few meters, but in severe conditions, they can be greater. If we know the weather conditions for the area in question, adjustments can be made to the images.

To compare the images from two different satellites, you have to know their height above the ground precisely – if your knowledge of the height of one of the satellites is in error by, say, 10 m, then that will appear as an elevation on the ground below of perhaps 20 m or more different from the truth.

Having the two radars on the same Shuttle and thus taking both images simultaneously got rid of the errors caused by the issues outlined above. The remaining errors are greatest in areas with steep terrain such as the Himalayas and the Andes, and very smooth sandy surfaces such as in the Sahara Desert. Figure 45 uses shading to illustrate the complete coverage achieved in South America – the same was true on other continents outside the polar regions. The full information is in digital form that can be processed directly by computers – the technical term for the information is a Digital Elevation Model (DEM).

The global coverage achieved by the Space Shuttle in 2000 provides a useful baseline for many purposes. It has been complemented recently by the release of an even more comprehensive DEM by NASA and the Japanese Government based on images taken by Japan's ASTER[16] camera on NASA's Terra satellite. Made available in 2009, this new elevation information comes from optical images taken by the ASTER stereo cameras – one looking directly down, the other facing backwards as the satellite sweeps across the sky. It took several years to collect the full set of reasonably cloud-free ASTER images compared with the 10 days for the Shuttle imaging radar, and then a year of computer processing to overlay the stereo images precisely, cut out any clouds and compute the elevations. The resulting DEM is still considered to be of "research quality" and thus may contain some faults, but it stretches from 83° north to 83° south, whereas the Shuttle data stopped at 56° (see Figure 46).

Satellites that can take high-resolution images, both optical and imaging radar, can generate elevation maps for specific areas that are more accurate and detailed than the Shuttle or ASTER versions. These maps may be required for flood prediction, planning for large-scale construction, flight simulation and many other applications. A commercial global DEM product with 2-m local accuracy and 12-m granularity is promised for 2013, derived from the TerraSAR-X and Tandem-X twin imaging radar satellites. For many parts of the world, the Shuttle and ASTER elevation information is still the best available and has the added benefit of being available free of charge.

---

[16]   ASTER = Advanced Spaceborne Thermal Emission and Reflection Radiometer.

Figure 45. Shading illustrates the elevation information of South America as measured by the Shuttle Radar Topography Mission on STS-99 in February 2000. Courtesy NASA/JPL-Caltech.

FARMING, FORESTRY AND LAND USE

During the Cold War, the CIA not only supported military coups in Chile and Iran, but it also estimated in advance the size of each year's grain crop in the Soviet Union. Unlike its clandestine military activities, the CIA's Soviet grain crop information was widely publicized. It provided the grain markets in the West with advance information about demand for Western grain by the Eastern Bloc countries

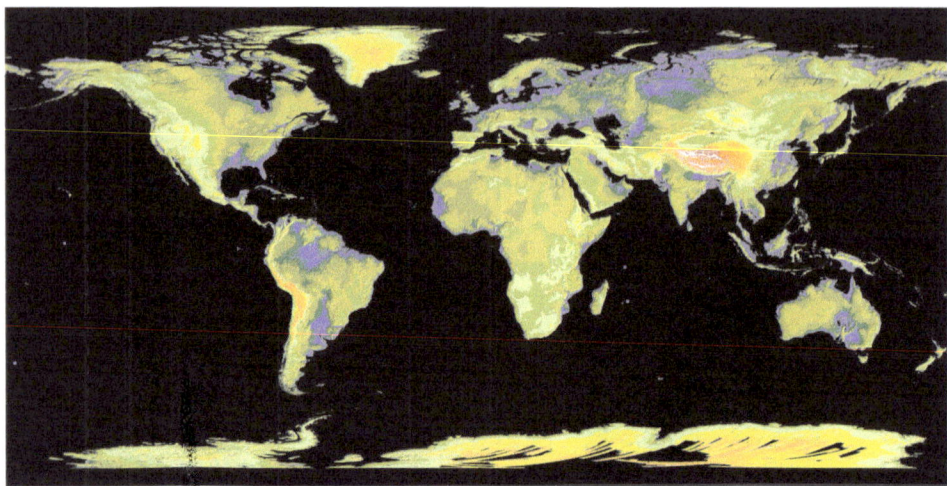

Figure 46. Color indicates elevation in this overview of the Digital Elevation Model (DEM) derived from ASTER images covering all of the world's land area, except the very heart of the Antarctic. Courtesy NASA/JPL-Caltech.

and, more generally, gave an indication of the health of the Soviet economy. From 1972 onwards, Landsat satellite images were acknowledged by the CIA as an important source of information for the grain crop forecasts. Images were taken at several points in the growing season across the Ukraine and other parts of the Soviet grain belt, allowing the analysts to estimate the size of the eventual harvest. In 1972, the CIA estimate suggested that the Soviets would be largely self-sufficient that year and thus would not need to import from the West. That particular estimate was high by more than 10% and in the event, the Soviets quietly bought 15 million tons of Western grain, pushing up prices in the West. From then on, the CIA spared no effort in using all available sources of information, high on the list of which were images from Landsat.

Ironically, the CIA estimates were widely read in the Soviet Union itself. According to Sergo Mikoyan, whose father was a Soviet Politburo member, "the Kremlin, the Council of Ministers, and the *Gosplan* (State Planning Committee) between the 1960s and the 1980s relied increasingly on CIA data for such important matters as the grain crop, rather than on reports from local Communist Party bosses". He says the reason the Kremlin relied on information from its sworn enemy was because the CIA "used satellite photos and therefore was able to publish its findings before Soviet authorities could even inform their leaders". He also cited the tendency of local Communist Party bosses to engage in "wishful thinking and a desire to favorably impress Kremlin authorities", which made their forecasts unreliable.[17]

---

[17]   Noren (2007), and Mikoyan (2001).

Thus, the use of satellite images for managing agricultural affairs has a long history. Today, the CIA's role has been taken on by the Foreign Agricultural Service of the US Department of Agriculture, which is mandated to provide "market intelligence in the form of timely, objective, unclassified, global crop condition and production estimates, for all major commodities, for all foreign countries". The estimates are based on crop production information from foreign government reports and field visits, but it is the comprehensive view afforded by space-based earth-observing satellites, such as Landsat, that provides "the unbiased, global, farm-level observations necessary to objectively verify these reports".[18] Due to a fault on Landsat-7 in May 2003, since then, the satellite images have come mainly from India's IRS satellites, topped up with some from France's SPOT and the Brazil/China CBERS satellites.

The story in Western Europe also has bizarre origins. In the late 1980s, olive plantation subsidy claims under Europe's Common Agricultural Policy (CAP) system by farmers on the Italian island of Sicily raised suspicions when the total olive-growing area approached the area of the whole island. Sicily is the home of the Mafia and inspectors sent by the authorities to check the subsidy claims were at risk of receiving a traditional Mafia welcome. The research unit of the European Union, based in northern Italy at Ispra in the foothills of the Alps, decided to try using satellite images to check the claims of the farmers. This non-invasive approach proved successful and evolved through the 1990s into the recommended and sometimes mandated method for verifying CAP claims.[19] Typically, the satellite photos are used to suggest areas where an on-the-spot inspection is required, although, in some cases, the satellite images alone are used as the basis for decision – some of the legal implications of this were discussed in Chapter 1.

The satellite images aren't infallible and there are many stories of attempts to deceive them, such as the painting on buildings of fake olive trees by Greek farmers to fool the satellites and boost subsidies.[20] But satellite images not only prevent inspectors being physically threatened; they also prevent inspectors or even governments from cheating – as was occasionally the accusation of one European country about another. Satellite images may not be perfect, but they are objective.

How can you tell the size of this year's wheat crop or whether a farmer is growing olives just by looking at satellite pictures? Some satellites take pictures in multiple colors, some of the colors being in the infrared, which the human eye can't see – we feel infrared light as heat but we can't see it. Each color is photographed separately but simultaneously – as explained in Chapter 1 for Japan's Daichi satellite. Computers on the ground then combine the colors in various ways so that items or topics of interest are highlighted.

Figures 47 and 48 illustrate the idea for a farming application. The location is northwest Minnesota on the border of North Dakota in the USA just north of Fargo and Moorehead, with the Buffalo River on the right merging into the Red River that

---

[18]   NASA website, *http://landsat.gsfc.nasa.gov/news/news-archive/soc_0010.html*.
[19]   Smith (2004).
[20]   See *www.euro-know.org/europages/dictionary/f.html*.

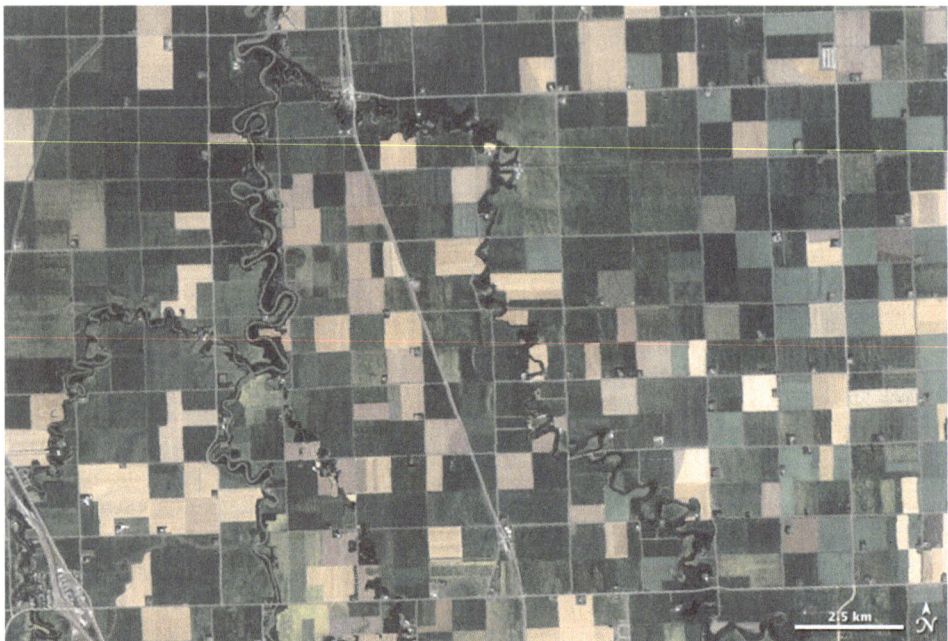

Figure 47. Farmland in Minnesota as viewed by Landsat-7 in true color. Credit: NASA's Earth Observatory.

Figure 48. Farmland in Minnesota as viewed by Landsat-7 in false color. Credit: NASA's Earth Observatory.

flows due north to Lake Winnipeg. The images were taken on September 10th 2009, near the end of the growing season – some crops have already been harvested, leaving squares of tan and brown. Figure 47 is in natural colors, more or less as you would see it with the naked eye from above. Figure 48 is a combination of three colors ("spectral bands" to give them their technical name), which allows various plant conditions to be seen. As the NASA website explains, "To the untrained eye, the false-colour images appear a hodge-podge of colours without any apparent purpose. But local organic farmer Noreen Thomas is now trained to see yellows where crops are infested, shades of red indicating crop health, black where flooding occurs, and brown where unwanted pesticides land on her chemical-free crops. The images help the Thomases root out problems caused by Canadian thistle and other weeds. They help confirm that their crops are growing at least 10 feet from the borders of a neighboring farm – required to maintain organic certification. They can also spot the telltale signs of bottlenecking in the fields – where flooding is over-saturating crops – and monitor the impact of hail storms. Just as remote imagery informs Thomas when it's best to rotate crops in her 1,200 acres, she can also determine when her cows need a new pasture. When the large herd of cows chews its way through the landscape, satellite images show where the cows may be overgrazing".[21]

The technique developed with the early Landsat satellites is to compare the scene as seen by the red camera with that as seen by the infrared camera that is just beyond red, namely just beyond what the human eye can see, the so-called "near infrared". The difference in brightness of a field between the red and the near infrared is a measure of the chlorophyll in the vegetation. By convention, this difference is usually shown in a false-color image as a shade of red – a deep, dark shade indicates a large difference and therefore strong growth of vegetation.

The appearance of the fields changes with the season and with the alternating plots of organic wheat, soybeans, corn, alfalfa, flax or hay. By taking images at several times during the growing season, it is possible to monitor the health of the crop and decide whether more fertilizer or weed killer is required – not only field by field, but within individual fields. The time sequence of images during the growing season makes it possible to predict the eventual yield at harvest time. To get an accurate yield forecast, information is needed about the irrigation of the crop, rainfall and other weather conditions. Meteorological information for the period since the crop was planted and a forecast of rain and temperature up to harvest time are combined with the crop status as seen in the satellite images. This allows the government agencies to predict a country's harvest for specific crops.

Images at different times also help to distinguish one crop from another. Not only does the speed of growth vary from crop to crop, but the color in the false-color image will vary in time characteristically for each crop. The European agencies managing farm subsidies exploit this technique to check farmers' subsidy claims.

---

[21] See *http://landsat.gsfc.nasa.gov/news/news-archive/soc_0022.html.*

France's SPOT satellites and India's IRS satellites have four spectral bands (colors) rather than seven and they, too, are used for farming applications. Germany's RapidEye fleet of five satellites each has five spectral bands and claims that farming is the premiere market it is addressing. One of the five spectral bands is what RapidEye calls "the Red-Edge band", which is right at the edge of what the human eye can see before it becomes invisible infrared. RapidEye considers that this color is particularly sensitive to changes in chlorophyll content and thus is useful for monitoring the health of vegetation, distinguishing crops more assuredly and helping to measure protein and nitrogen content in biomass. The US commercial WorldView-2 satellite also takes Red-Edge images but with 2-m resolution compared with RapidEye's 6½-m. Jill Smith, whose DigitalGlobe company owns WorldView-2, says this feature allows analysts "to identify more varieties of vegetation, [distinguish] cotton-based camouflage from natural ground-cover or track coastal changes". DigitalGlobe's advertising says that the Red-Edge feature allows the farmer to "discriminate young vs. mature plants, conifers vs. broad leafed plants and even detect subtle changes in plant health, before they are visible".[22]

One of the distinguishing features about RapidEye is that it consists of five identical satellites to improve the chances of getting cloud-free images and thus ensure rapid (hence the name) revisits to a scene during the agricultural growing season. The RapidEye satellites weigh just 150 kg each and cost $230 million in total to build and launch all five into orbit.[23]

India claims to have the "largest civilian remote sensing satellite constellation in the world".[24] The fleet of 11 working satellites (as of July 2010) is mostly in the ⅔–1½-ton class, made up of a mix of satellites capable of taking stereo high-resolution images (Cartosat-1, -2, -2A and -2B), which are of particular interest to the Indian military, low-resolution ocean monitoring images (Oceansat-1 and -2), radar images (Risat-2 weighing 300 kg), images with low cost or experimental technology (TES and the tiny IMS-1 weighing 83 kg) and dedicated resource-monitoring satellites (Resourcesat-1 and IRS-1D). Only the last two are really equipped to provide agricultural information, outnumbered by RapidEye's five. RapidEye's Chief Executive Officer, Wolfgang Biedermann, believes that the reliable revisit ability of his fleet of satellites "will provide the impetus to establish new types of businesses which depend on the high frequency and reliability of data". The 6½-m resolution of the RapidEye images is competitive with the images from SPOT and IRS – and better than Landsat's – but it is the ability to take frequent cloud-free pictures of the same area that Biedermann stresses.[25]

DigitalGlobe takes a different approach, being able to store a large volume of images in the satellite – over 2,000 gigabytes (GB) in WorldView-2, which is more

---

[22] See *www.rapideye.de/home/system/satellites/#camera*; Longhorn (2010a); *www.digitalglobe.com/downloads/spacecraft/VegAnalysis-DS-VEG.pdf*.
[23] Taverna (2008).
[24] See *www.isro.org/scripts/currentprogrammein.aspx#IRS*.
[25] Taverna (2008).

than the capacity of the hard disk in most home computers. DigitalGlobe President Jill Smith says this allows the company to rapidly build up an enormous image library so "that for our customers more than ever the imagery they want is already available". The WorldView-2 satellite was launched in October 2009, joining World-View-1 and Quickbird in the DigitalGlobe fleet. These three satellites illustrate the steady improvement in commercial surveillance satellites in the past decade. Quickbird was launched in 2001 and offers black-and-white images with 61-cm resolution or color images with 2½-m, and has an image memory of 128 GB. Launched 6 years later, WorldView-1 has the expanded image memory of 2,000 GB mentioned above and black-and-white images with 50-cm resolution. Both of these satellites are less than 500 km high in order to provide their detailed imagery, but WorldView-2 is nearly 800 km up and still provides black-and-white images with 46-cm resolution – or color images with 2-m resolution, as mentioned previously. The higher altitude of WorldView-2 gives it a wider field of view.[26]

Commercial companies now offer services to particular farming groups, not only for major crops such as wheat, corn and rapeseed, but also for smaller crops such as to help French wine-growers improve the health and yield of their grapes. One French satellite imaging company, Infoterra, claims that 10,000 farmers in France use their satellite image-based service. A UK precision farming company, SOYL, provides statistics for the improvement satellite images can make via its SOYLSense information service. One of their clients, farm manager Simon Beddows, notes that "the service lets you apply exactly the right amount of nitrogen the crop needs. We got an extra 0.45 ton per hectare from using SOYLSense on one trial and almost an extra ton per hectare from the second trial – that's a yield benefit of 4% and 8%". Another supplier is the Loris service of Finland's Kemira. Their advertising claims that "the fields which benefit most from variable rate nitrogen are those with the greatest inherent variability [where] yield benefits over 10% have been consistently recorded".[27]

Manufacturers of farm equipment now integrate satellite imagery and satellite navigation into their products. Modern tractors used for spreading fertilizer or weed- or pest-killer can be programmed to dispense the appropriate amount of the additive based on crop health as seen in the satellite images – the tractor's satnav tells its computer where it is relative to the satellite image. The farmer inserts a memory stick into a slot in the dashboard that loads a map so that the fertilizer or herbicide or pesticide is automatically dispensed as the vehicle is driven. The savings in fertilizer and weed/pest-killer make economic sense, and the reduced run-off of these substances into the groundwater is an added environmental benefit.

Better harvests aren't the only reason to use satellite images. In the UK, British Sugar hired my employer, Logica, to forecast the yield of the sugar beet crop. By

---

[26] Longhorn (2010a); DigitalGlobe website *www.digitalglobe.com/*.
[27] Musquère (2009), and Longhorn (2010b); *www.infoterra.fr/Infoterra*; *www.stackyard.com/news/2006/12/arable/01_kemira_growhow.html*.

knowing the size of the harvest 2 or 3 months in advance, British Sugar could order the appropriate quantity of ancillary materials, transport and staff, which led to overall business efficiencies. We purchased satellite images of the beet-growing parts of England four or five times during the growing season, automatically identified the fields growing beet and analyzed the health of the crop in each, and then combined this with weather information to estimate the yield at harvest time. In another project at Logica, we monitored all of the UK's potato fields, not to estimate yield, but to verify the acreage reported by farmers, because the annual subscription of a potato farmer to their trade body (the Potato Marketing Board) was based on acreage – the bigger the acreage, the bigger the fee. By digitizing all of the acreage reports, integrating them with digital maps of the UK's field boundaries and then examining satellite images to see whether any unreported fields contained potatoes, a list of anomalies could be prepared for field inspectors to check on the spot. This enabled the Potato Marketing Board to reduce the number of field inspectors and yet increase the effectiveness of policing the annual subscription. The total cost of the satellite analysis including buying the necessary images from commercial satellite operators was a few tens of thousands of dollars – about the cost of a single field inspector – and the level of unreported acreage plummeted as farmers learned about the "spy in the sky" that was checking on their reported acreage. The immediate benefit was to increase the subscriptions received by the Potato Marketing Board, but in the medium term, they were able to reduce the fee per acre due to less fraud and cheaper enforcement.

Satellite images are central in attempts to monitor the world's opium crop. RapidEye, for example, mapped the opium-growing regions of southern Afghanistan, an area the size of Oregon or the UK, in less than 2 weeks. That map characterized the vegetation across the region and is then used as a reference for the rest of the growing season – highlighting areas most likely to produce the crop.[28]

The world's forests are recognized as being critical to a sustainable climate and satellite images are the only global and objective way to monitor them – this is not farming as such, but since trees are a form of vegetation, it seems reasonable to discuss them here.

Deforestation accounts for about a sixth of all global greenhouse gas emissions, which is significantly more than the whole global transport sector.[29] In addition, tropical forests are home to over half of the world's species of plants and animals, and are thus a particular concern as a reservoir of biological diversity. Britain's forestry over-lord Tim Rollinson sums up the issue: "Deforestation might yield land for other purposes but it destroys bio-diversity and increases erosion and flooding." Forestry expert Sir David Read, whose authoritative 2009 "Read Report" on the subject was sponsored by Rollinson's Forestry Commission, quips that "Forests are not only nice places to walk". He then points out that "they're also profoundly

---

[28]   Parmalee (2010).
[29]   Pachauri and Reisinger (eds) (2007) p. 14.

important to carbon sequestration, a process that remains one of our best hopes for long-term climate rebalancing". The IPCC says that "forests sequester the largest fraction of terrestrial ecosystem carbon stocks" more than the oceans, lakes, wetlands and farmland combined! About half the weight of a tree is carbon, so the bigger they grow and the more of them there are, the less carbon there is to make carbon dioxide and the other greenhouse gases. Plus, unlike crops, they don't need fertilizer or weed- or pest-killer, thus saving on the raw materials and energy to make those chemicals and avoiding groundwater pollution from the run-off.[30]

The Brazilian space agency (INPE) uses satellite imagery to improve the annual estimate of Amazon basin deforestation. It uses images from its own satellites, CBERS-1 and CBERS-2, but the persistent cloud cover dictates that it buys images from several other satellite operators, too. The six[31] satellites in the Disaster Monitoring Constellation (DMC) discussed in Chapter 5 take 600-km-wide images of roughly Landsat-type quality and have been contracted since 2005 to observe the Brazilian forests (Figure 49). By comparison, Brazil's own CBERS satellites produce images of similar quality just 113 km wide and have suffered some failures in orbit.[32] By virtue of having multiple satellites, each with a wide field of view, DMC ensures full imaging of the cloud-prone Amazon basin. Radar images can see through the clouds but detecting breaks in the forest under the clouds is difficult because the radar images are subtly different from ordinary optical ones, as discussed below.

There is concern that illegal loggers in the Amazon are becoming wise to the ability of satellites to detect their activities. Most of the satellites used by Brazil until now had a resolution of about 30 m – that is to say, attempts to zoom in to see objects smaller than 30 m gave a blurred image. The lumberjacks are said to have realized this and began to thin out the forest, leaving enough of the forest canopy to fool the satellite into thinking that the forest is untouched. Given the lack of zoom capability of the satellite images, the authorities may assume that the areas in question are untouched. DMCii (the company that sells DMC images) Chief Executive Dave Hodgson says that the optical images DMC provides can detect a thinning of the forest canopy, not just its total removal. Despite his confidence, DMC and others are undertaking research to check whether the images can be fooled. If that proves to be the case, one solution will be to use more detailed satellite images, which can detect each individual tree – DMC images blur several trees into each pixel, thus raising the possibility that loss of some of the trees might not be noticed. Recently, Brazil agreed to place a better-quality telescope and camera on its next satellite – with the British-supplied instrument, Brazilian authorities will get images of 2–3-m resolution, allowing them to spot the removal of individual trees and thus detect any thinning out beyond question.

Using more detailed satellite images will cause two problems. First, each image will cover a smaller area – perhaps only a tenth or less of the 600-km DMC width. So

---

[30]   Fischlin *et al.* (2007) p. 214, and Bawa (2010).
[31]   Two more are due to be launched in 2010.
[32]   Stryker (2009) p. 21.

Figure 49. Farming development in the Amazon basin, Brazil, August 2005. Credit: DMC International Imaging Ltd.

Figure 50. An area of about 100 km on a side in western Brazil near the Bolivian border in 1984; the red areas are forest, the pale areas cleared land. © Fugro NPA Ltd, *www.fugro-npa.com.*

lots more images will have to be taken and processed. Second, the cost of the detailed images is usually high compared to DMC. In practice, DMC and other low-resolution (blurred) images may be used for most of the Amazon and high-resolution images purchased only for those areas that appear suspect.

Two more images of the Amazon basin are Figures 50 and 51, separated in time by 17 years. "Deforestation of jungles in Brazil is dramatically exposed by [these] two images," said the explorer Sir Ranulph Fiennes and he noted that "some 15,000 sq km/5,800 sq miles are being slashed and burned every year".[33]

As Figure 52 shows, Brazil is by far the biggest contributor to world deforestation. The good news is that satellite imagery has underpinned a successful campaign by the Brazilian Government to deter illegal forest clearance (see Figure 53). The key seems to be that the satellite images are used not just by the Brazilian space agency, but have become a routine tool for the local government agencies responsible for protecting the forest.[34]

The generally disappointing Copenhagen climate conference in late 2009 did at least one good thing in recognizing the importance of an initiative called REDD-plus (or REDD +) – Reducing Emissions from Deforestation and forest Degradation in developing countries.[35] The idea is to pay poorer countries to preserve forests through conservation and sustainable management. The USA and five other wealthy countries[36] have put their money where their mouth is and kick-started the initiative with $3.5 billion. REDD + will require the recipient countries to have an operational forest-monitoring program and satellite imaging is an integral part of any such scheme. Satellite imagery will check a country's forests several times a year and direct ground-enforcement personnel to areas requiring attention – frequent imaging is important because after a year, degradation is hard to detect.

Returning to the subject of farming with which we started this section, satellites can help measure another of the world's non-renewable resources – topsoil. In most areas, it takes centuries to generate an inch of topsoil and it is being eroded faster than new soil forms in a third of the world's cropland. If the vegetation that binds the soil disappears, wind and rain will rapidly sweep it away. Roads, buildings and other non-farm uses of land can aggravate the loss of farmland by interrupting natural drainage and irrigation patterns – concreted car-parks and home driveways can increase run-off so that topsoil is threatened. Satellites can highlight and monitor the issue but political will is needed to regulate land use and prevent the loss of this critical asset.[37]

---

[33]   Fiennes (2006) pp. 6–7, 260–261.
[34]   Roy Gibson, private communication.
[35]   United Nations (2009) p. 2.
[36]   Australia, France, Japan, Norway and the UK.
[37]   Brown (2009).

Figure 51. The same area as that shown in Figure 50 in 2001; the fishbone pattern of forest clearance ties in with the creation of roads and subsidiary roads that give the loggers access to the forest. © Fugro NPA Ltd, *www.fugro-npa.com.*

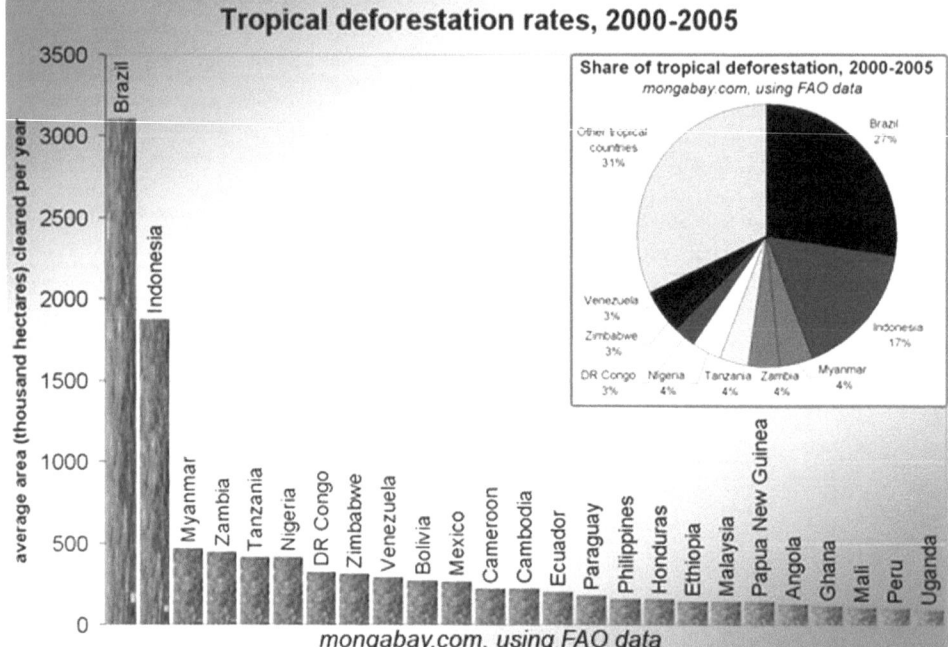

Figure 52. Tropical deforestation around the world – Brazil and Indonesia together account for almost half. Credit: Rhett A. Butler/mongabay.com.

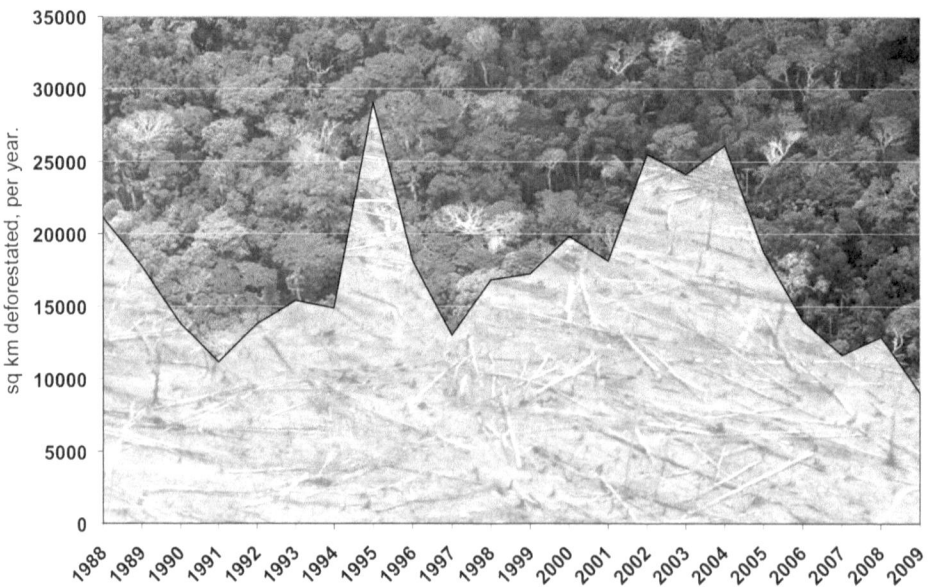

Figure 53. Reduction in Brazil's deforestation rate since 2004. Credit: Rhett A. Butler/ mongabay.com.

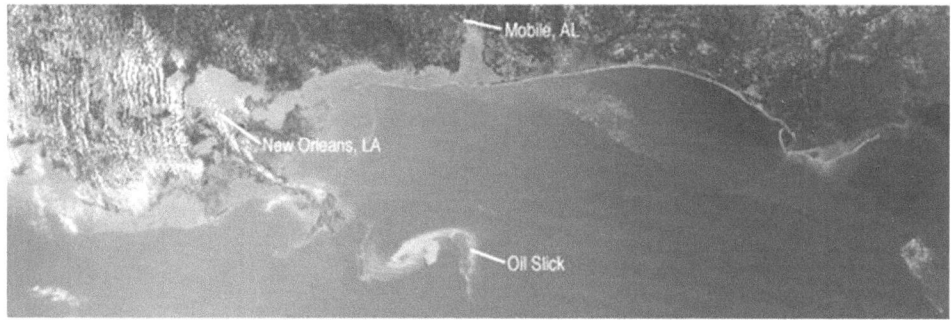

Figure 54. An oil slick off the Mississippi delta in the Gulf of Mexico. The top image taken by NASA's Aqua satellite gives an overview of the extent of the slick 9 days after the April 20th 2010 blowout as it approached the coast, while the lower image taken by NASA's EO-1 satellite is a close-up of the edge of the slick on April 25th as the attempts to contain the oil got underway. Credit: NASA/Earth Observatory.

OTHER TYPES OF INFRINGEMENT

In addition to farming and forestry discussed above, other enforcement activities in which satellite imagery is increasingly the norm include oil spills at sea and fishery conservation. In many cases, the detection is a case of watching for change to occur in an otherwise static scene. Image-analysis companies now offer a "change alert" service whereby they contact you if a change has occurred in an area you specified (you can specify what sort of change you are interested in) and send you a thumbnail version of the satellite image so that you can decide whether to buy the full image.[38]

---

[38] See, e.g. the SPOTMonitoring service at *www.spotimage.com.*

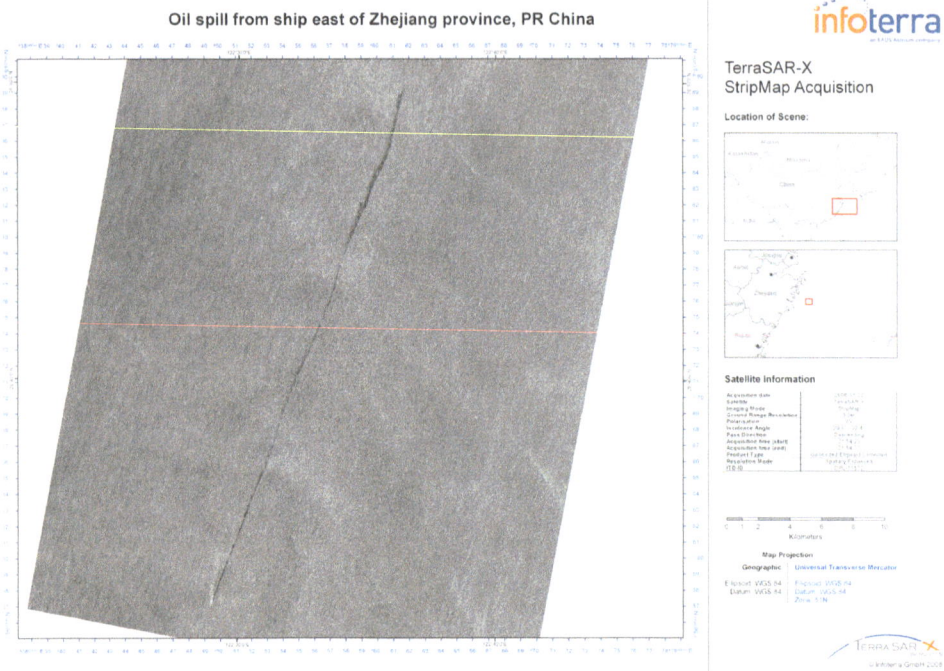

Figure 55. Oil trail clearly visible in this TerraSAR-X radar image taken on November 22nd 2008. Credit: Infoterra GmbH.

Oil spills show up in both radar and conventional images because of the smoothing effect the oil has on the sea surface.[39] Figure 54 shows two satellite views of the BP oil spill in April 2010 resulting from the explosion on the Deepwater Horizon offshore drilling rig in the Gulf of Mexico that killed 11 people and injured scores more. In the lower image, ships can be seen working at the edge of the slick and the individual waves are also visible. The upper image gives a bird's eye view of the slick as a whole and its proximity to New Orleans, LA, and Mobile, AL. In this case, the oil slick was known before the satellite images were taken. In other cases, the sequence is reversed, with a satellite, usually a radar imaging satellite, spotting the tell-tale track of an oil slick, often with a ship discernible at one end of it. An enforcement vessel can then be dispatched to intercept the culprit. An example of an oil slick detected by imaging radar is shown in Figure 55.

Radar images give considerable information about water and the things on it and under it. If the sea or lake is calm, it looks dark in a radar image. The satellite transmits its radio signal off to the side, which, if the earth's surface is bumpy, will be reflected back towards the satellite by the bumps. But if the ground is perfectly flat (as a mill pond, so to speak) and therefore lacks bumps, the radio signal is reflected

---

[39]   John Amos's website *www.skytruth.org* provides more information on this subject.

away from the satellite as if by a mirror. Thus, no signal comes back to the satellite and the resulting image is black. In flat seas, therefore, any ripples or disturbances stand out clearly in a radar image. Figure 55 illustrates this in reverse. The sea off the east coast of China is a bit choppy, but the oil being leaked by a ship calms the sea (hence the phrase "to pour oil on troubled water") and leaves a black trail in the image.

Reduced fish stocks across the globe are leading to national and international regulations to control over-fishing. The European Union requires all fishing vessels over 24 m in length to switch to electronic logbooks by the end of 2009 and those over 15 m long by June 2011 (about 10,000 vessels). They have to report their catch at least every 24 h while at sea. Among the benefits of the system is the elimination of problems with hard-to-read handwriting. An additional feature being discussed is to have satellite links that would allow regulators to capture nearly real-time information on the quantity of fish being caught by each country and help ensure stocks are harvested effectively. Rapid feedback from the authorities via a satellite link on each report would help fishermen adapt their actions to maximize their catch and still stay within the regulations, instead of having to stop at 90–95% to retain a safety margin.[40]

Three-quarters of all marine fish species are at the brink of, or actually below, sustainable population levels. This situation is a horrific example of the ironic sequence of events known as the "tragedy of the commons" in which short-term gain by the individual leads to over-exploiting a common resource to the long-term detriment of all, including themselves. Marine Conservation Professor Callum Roberts at the University of York in England argues that 10–20% of the ocean (especially feeding grounds and migration routes) should be no-take zones or marine reserves – currently, they occupy less than 1% of the world's oceans. Conservation expert Tristram Stuart explains that "these reserves become spawning grounds and nurseries for stocks, thus increasing the number of fish available for fishermen to catch beyond the borders of the reserve". Monitoring whether any vessels are in a no-take zone is intrinsically easier to enforce than quota systems that are getting more and more complex. For example, in Europe, the quotas are replete with definitions of the type of allowable fishing gear, limits on discarding fish, banning of "high-grading" in which fish of a legal landing size are thrown away to make room for larger more valuable fish and so on. Electronic logbooks would facilitate a switch from quotas to no-fish zones, as would the "change alert" service mentioned above. Unlike the current complex, fraud-prone and wasteful quota-based systems in use around the world, a no-fishing zone is easily policed from space. If there's a ship in the zone, it's illegal, with perhaps special forms of identification for research vessels or scuba-diving tourists.

Public awareness of discards by fishermen is high in the USA. Alaska, for example, bans it and requires "by-catch avoidance technology" – "by-catch" is

---

[40]   Inmarsat (2010) pp. 52–55.

anything a fisherman didn't intend to catch. Despite this, according to Tristram Stuart, the Gulf of Mexico shrimp trawl fishery has about the worst discard record of any individual fishery in the world, throwing back nearly half a million tons of snappers, emperors, turtles, etc.[41] The horrendous oil spill shown in Figure 54 is interfering with the activities of these shrimp boats, but that is little consolation for the region's fish stocks, which are being poisoned by the millions of gallons of oil gushing from the leaking oil well.

All ships over 300 tons are required under international regulations to carry an Automatic Identification System (AIS). This is a black box that sends out messages about the identity of the vessel, its whereabouts and its contents every 6 s. By comparing a map of all such ships with a satellite image that shows all ships in the region, illegal vessels can be spotted. These may be smuggling illegal immigrants or contraband between countries. They may be pirates off the African Coast. They may be fishing illegally. Once an unidentified vessel is spotted, surface or airborne inspection can follow. In another scenario, satellite imagery may spot a ship leaking oil (deliberately or accidentally) as it sails along. Tying an AIS identifier to the ship in the image gives the authorities what they need to pursue the culprit. Satellites can pick up the AIS signals far from land and several companies offer tracking services that rely on this. Virginia-based entrepreneur Dino Laurenzi launched two tiny satellites in 2009 dedicated to this purpose and two more of the 13-kg midgets are due to be launched in 2010. His company, SpaceQuest, is targeting not only the security services, but also commodity traders and other commercial companies. The first two satellites pick up half a million signals from 22,000 ships each day. That information is valuable to "commodity traders, who are very interested in knowing about the flow of energy around the world," says Laurenzi. Other satellite owners collecting AIS signals include London-based Inmarsat and New Jersey-based Orbcomm.[42]

The European Union is trialing a service to combine satellite imagery and AIS information on a routine basis. Imaging radar satellites provide imagery of the Mediterranean and other areas selected for the trials. All detected ships are then compared with the database of AIS reports and suspect vessels identified. This information is then passed to the Coast Guard and other security authorities for action.[43]

MINERAL EXPLORATION

Figures 56, 57 and 58 illustrate the wealth of geological information that can be obtained from false color satellite imagery. As discussed above, many satellites

---

[41]   Stuart (2009) pp. 127–135.
[42]   Brinton (2010f).
[43]   Land and Sea Monitoring for Environment and Security (LIMES); see *www.fp6-limes.eu/index.php*.

Figure 56. Part of the Hammersley Range in the north of Western Australia, about 120 km from north to south. © Fugro NPA Ltd, *www.fugro-npa.com.*

contain special cameras that photograph the scene below in several colors (or, more accurately, "spectral bands"), storing each color as a separate image. Computer manipulation on the ground is then used to combine the various colors to give a "false color" image, choosing the colors to highlight features of interest. In the previous section, we showed false color images tailored to show information relevant to farming and forestry. Other spectral combinations show up geological features.

Figure 56 uses green coloring to highlight the iron-rich rocks in the Hammersley Range in Western Australia. Iron is widely mined in this region, mostly in open pits; one mine is evident in the lower right of the image as a red stain on the lower right of the green circular feature. Hematite is the principal iron mineral seen in banded iron formations of the heavily eroded Precambrian-age mountains.

Figure 57 shows the intricate saw-tooth geology in an area formed by the collision of the Arabian tectonic plate on the left with the Eurasian plate. The area is in the southwest of Iran on the Persian Gulf. Just visible in the right of the image are the Zagros Mountains, which extend far to the north and east. The area near the coast contains major natural gas fields. Until satellite imagery became available, it was difficult to appreciate the large-scale nature of these geological features. Large-scale geological maps showing the fault and slip structure of a region can be drawn up from images such as this. In some inaccessible parts of the globe, this is virtually impossible to perform any other way – inaccessible because the conditions are extreme, such as in polar regions, or the politics are unfriendly, or the danger of personal injury or death is too high (in a failed state, for example). In such cases, the satellite images enable the planning to be performed before having to commit

Figure 57. Saw-tooth-like anticline (left) parallel to the Persian Gulf in southwest Iran with the river Mand in the lower right (about 70 km from north to south). © Fugro NPA Ltd, *www.fugro-npa.com.*

Figure 58. Tehran, in the lower left of this false-color image, with the Elburz Mountains to its north (about 120 km from north to south). © Fugro NPA Ltd, *www.fugro-npa.com.*

personnel to on-site activities such as seismic surveys, test drilling or in-field sampling, and thus minimize the time spent actually in the region.

The capital of Iran, Tehran, is about 800 km north of the previous image and Figure 58 shows how it is hemmed in to the north by the Elburz Mountains. Chromium is one of the minerals mined in the area. The peak of Mount Damavand can just be seen on the center-right edge of the image – the highest peak in the Middle East and the highest volcano in Asia, at 5,610 m (18,406 ft).

As we have already seen, satellite images, especially radar images, to be discussed below, are used to detect oil spillage at sea – both wide-area oil spills, such as following a tanker collision or sinking, and smaller-scale events, such as when a ship cleans out its tanks. One man's pollution is another man's gold dust and, in this case, satellite images of oil seepage at sea are of interest to oil exploration companies. Undersea oil reserves are frequently detected by virtue of the oil seeping through to the sea floor and thence to the surface.

CONSTRUCTION

Major construction projects nowadays involve computer-based mapping of the environment, for which one of the sources of information is satellite imagery. Whether the project is a power station, a bridge, a road, an oil refinery, a factory, a retail complex, a transport hub or a residential complex, detailed mapping is required not only for the architectural and construction teams, but also for such things as environmental impact analysis, insurance analysis (floods, earthquakes, subsidence, etc.) and the financial risk analysis.

The Chinese authorities, for example, used regular satellite images of the Beijing Olympic complex to provide clear and objective evidence to the International Olympic Committee that progress on the construction of the 2008 Olympic facilities was on schedule.

Satellite images come into their own where the terrain is inaccessible, inhospitable, poorly mapped or stretches over large distances. Building an oil pipeline in West Asia would be an example of this. The pipeline is likely to cross one or more earthquake-prone areas and may well span a mountain range or two as well as crossing national borders and perhaps areas of civil unrest. Both large-scale and detailed local maps will be required to plan the route and in poorly mapped areas, that is likely to involve satellite imagery and probably satellite-derived elevation information as discussed earlier in this chapter.

Even in well mapped cities, satellites can shed light on risks such as subsidence. Figure 59 shows the change in ground level in central London over an 8-year period as measured by imaging radar satellites. Two lines of previously unappreciated subsidence are shown in red. The first, in the upper half of the image, has occurred where a new underground train (Tube) line was being built – the $5 billion Jubilee Line extension – with the subsidence running from Pall Mall on the west of the Thames to London Bridge. The second, between Battersea Park and Clapham on the lower left edge of the image, is not as clearly associated with a single major construction project.

RECENT NEW FORMS OF SATELLITE IMAGING

In Chapter 8, imaging radar, hyper-spectral imaging and other advanced and even

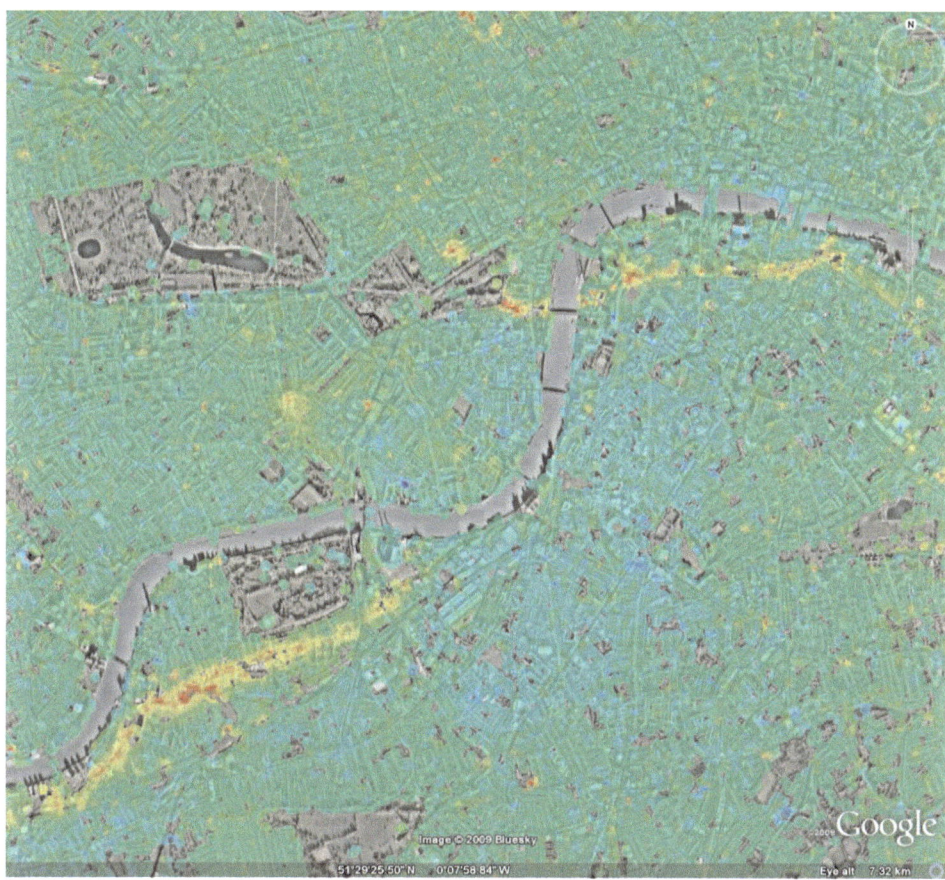

Figure 59. Ground motion in central London is shown in this map derived from radar images taken by the ERS and Envisat satellites between 1992 and 2000; red indicates subsidence, green stability and blue uplift. Hyde Park (leftmost) and Green Park/St James's Park are visible as gray areas to the left of the Thames, which winds from bottom left to upper right; Battersea Park is just below the Thames and Clapham Common is at the lower edge of the image. © Fugro NPA Ltd, *www.fugra-npa.com.*

speculative forms of satellite imagery will be discussed. Many of these forms of imaging are of interest in the commercial world, although it has to be said that the military remain the most enthusiastic customers for the moment.

The earliest imaging satellites (all military) found that 50% of the images they took were cloud-covered. The best that could be done to improve the ratio of cloud-free to cloudy images was to use weather satellites to try to warn the imaging satellite when the area of interest was cloudy – and if it stayed cloudy for a long time, there was nothing you could do.

Another problem for the first imaging satellites was that they only obtained good images when the sun was shining – they didn't work in the dark. Taking images in the infrared gives some degree of nighttime imaging but with much less contrast and

Figure 60. COSMO-SkyMed Image. © ASI (2009). All rights reserved.

somewhat lower resolution. Satellites continue to take infrared images at night, but they are not an alternative to conventional optical imaging, but more a complement to provide information on specific features such as heat emission. And infrared doesn't see through clouds, so you still have that problem.

The answer was recognized to be imaging radar, in which the satellite transmits a radio signal and detects the echo from each point below, thereby building up an image. A lot of processing is required to separate out the individual echoes and it wasn't until the 1970s that it became technically feasible and just about affordable to build a satellite with the necessary power and sophistication. The high price of the technology meant that military and scientific users were the only communities willing to finance these satellites until the end of the 20th century. Now, in the first decade of the new century, the improvement in electronics has dropped the price of imaging radar satellites to the point at which commercial companies have been persuaded to build and launch them. Table 5 in Chapter 8 (military) lists the main examples of these satellites.

A radar image of farmland shows up individual fields (see Figure 60) and, over time, changes in the field can be seen. However, a lot of research is still needed to reach the point at which radar images give us as much information as optical images, especially the multi-spectral and hyper-spectral ones.

The US military has more experience of radar imaging from space that anyone else, so other characteristics of radar images are covered in the discussion on military surveillance in Chapter 8.

5

Society and survival

To fully appreciate the contribution of satellites, we must look beyond commerce
and defense. A comprehensive treatment of this subject could fill a volume, for
example to describe the scientific research being undertaken using satellites –
attempting to measure the changing earth in new ways and in more detail. In this
chapter, we look at two important ways in which surveillance satellites save lives –
helping to deal with natural disasters and to conserve fresh water. As in other roles
for satellites, the key is their ability to span continents, to provide objective
information and to be independent of ground facilities such as electricity grids and
telephone networks.

NATURAL DISASTERS

The horror of the tidal wave that engulfed Indian Ocean coasts on December 26th
2004 is hard to forget. Television showed pictures taken on personal camcorders that
emphasized the stark contrast between the idyllic vacation setting of the tropical
beaches and the inexorable wall of water that swept all before it. Coming the day
after Christmas Day heightened the shock.

The waters swept away roads, telegraph poles and mobile phone masts as well as
homes, shops, schools, hospitals, farms, fields, cars and, of course, people. The death
toll was in the region of 200,000 people, especially in the Indonesian province closest
to the source of the tidal wave, but also in Thailand, Sri Lanka and other Indian
Ocean states.

It was difficult for relief agencies to know where to begin – all communications
with the area were down, except by satellite phone, and maps of the area were no
longer much use, given the scale of the devastation. Satellite pictures provided the
first evidence of the breadth of the disaster and indicated where roads still existed.

The first publicly available satellite images came from a group of satellites called
the Disaster Monitoring Constellation (DMC). These five satellites were the
brainchild of an energetic and brilliant British professor, Sir Martin Sweeting (he
was knighted by the Queen in 2002 for his endeavors). In the 1980s, in parallel with

P. Norris, *Watching Earth from Space,* Springer Praxis Books,
DOI 10.1007/978-1-4419-6938-5_5, © Springer Science+Business Media, LLC 2010

his duties as an engineering lecturer at the University of Surrey near London, Sir Martin began to design and build small satellites, weighing typically 50–70 kg (100–150 lbs). The satellites were initially to demonstrate engineering technology and were sufficiently small and light to be carried into orbit in the spare space beside a big conventional satellite. Such "piggy-back" launches had the advantage of being relatively cheap, sometimes even free. Sir Martin expanded the scope of the small satellite business by helping universities and institutes in developing countries to build their own satellites. The engineers would come from their home country and work alongside Sir Martin's team in Surrey to help build their satellite, then return to Nigeria, Algeria, Pakistan or wherever having acquired skills and experience for the future.

Over the years, as technology improved, the small satellites produced by Surrey Satellite Technology Limited (universally referred to as Surrey Satellite) became capable of performing useful functions. By the mid 1990s, having launched a dozen or so satellites already, Sir Martin realized that a small satellite could now carry a camera and telescope that would have sufficient power to provide useful pictures. He recognized that he couldn't compete with the very detailed pictures taken by the established commercial and government earth observation satellites, so he decided to address the problem of getting pictures rapidly. Geostationary satellites can take pictures continuously but as noted in Chapter 3, these satellites are so far from earth (36,000 km) that they provide very little detail. Commercial earth observation satellites are in orbit a few hundred miles high in order to get detailed images, but that means that they move rapidly across the sky, returning to the same part of the globe days or even weeks later. Sir Martin reasoned that by placing several satellites in orbit each with a wide field of view, he could guarantee to get an image of anywhere in the world within 24 h. He realized that this relatively rapid response imaging would be especially useful in the aftermath of natural disasters, so he called the concept the Disaster Monitoring Constellation (DMC). He drummed up support for the idea in a few of the countries with which Surrey Satellite had developed ties and began to manufacture the first satellite.

With other earth observation satellites, it takes 3–5 days to get imagery of a new disaster area, so the 1-day response time of DMC is an important improvement.

Although its headline objective is to support the logistics of disaster relief, DMC's "day job" is to provide independent daily imaging capability to the partner nations Algeria, China, Nigeria, Spain, Turkey and the UK. Each country operates its own satellite and collaborates with the others through a UK-based coordination organization called DMCii. The satellites can take pictures that are 650 km wide and thus cover vast areas quickly. Each country has agreed to make up to 5% of the imagery available for disaster relief, although, thankfully, that limit has never been reached. When a disaster strikes, the main challenge is not the 5% limit, but whether the satellites are already scheduled to take images of other parts of the world. In principle, each satellite could take pictures continuously, since sunlight provides the electricity needed to keep going. The images have to be sent to the ground, of course, and there are only a few stations on the ground that can receive the images. The answer is to store the images in the satellite's memory until it passes over a station, so, in practice, the limit is the size of the satellite's memory. Once on the ground, the images are channeled to aid agencies initially through Reuters AlertNet.

Reuters AlertNet is a free service run by the Thomson Reuters Foundation – a charity set up in 1982 by the Reuters news agency. AlertNet is dedicated to providing information to humanitarian organizations and the public about disasters. The service was started after the Rwanda crisis in 1994 when poor coordination between aid agencies was recognized as a barrier to providing aid.

As well as benefitting from AlertNet, the DMC Consortium is also an active participant in another free service aimed at helping the response to disasters – the International Charter for Space in Major Disasters. The UN triggered the creation of this Charter by organizing the UNISPACE-III conference in 1999, with the ambitious aim of "creating a blueprint for the peaceful uses of outer space in the 21st century". Following the conference, held in Vienna (Austria), the European and French space agencies (ESA and CNES) made a commitment to set up and manage the proposed International Charter for "Space and Major Disasters".[1] Since then, 10 other countries have joined (in order of signing): Canada, India and USA, Argentina, Japan, UK,* Algeria,* Nigeria* and Turkey,* and China. The US membership covers the Weather Agency (NOAA) and the US Geological Survey, which, in turn, brings in two US commercial satellite organizations, Digital Globe and GeoEye. Each member agency has committed resources to support the provisions of the Charter. A Charter member can call a single number to request the mobilization of the space and associated ground resources of the member agencies to obtain data and information on a disaster occurrence. An Emergency On-Call Officer analyzes the request and the scope of the disaster with the member and prepares an archive and acquisition plan using available space resources.

The Charter was invoked on December 26th 2004, after the tsunami in south Asia. It covered three disaster-stricken zones along the coasts of southern India, Sri Lanka and Indonesia–Thailand. Member agencies contributed an estimated 200 satellite images, providing essential information about areas in which surface access was impossible or difficult. Images were used as base maps for assessing damage to infrastructure and coastal habitations, for measuring the extent of sea surge and for identifying the areas where emergency aid was most needed.

UK Minister for Science and Innovation Ian Pearson said "By working together, the Charter's international partners are helping to save lives across the globe". His successor as Minister, Paul Drayson, called it a "unique international partnership whereby the saving of lives takes priority over politics, and country borders become meaningless in the face of disaster".

The Charter has been activated over 200 times since it became operational in 2002 and is currently called upon about three times a month (see Figure 61).

---

[1]   See that organisation's website at *www.disasterscharter.org*.
*   Through the Disaster Monitoring Constellation (DMC) organization.

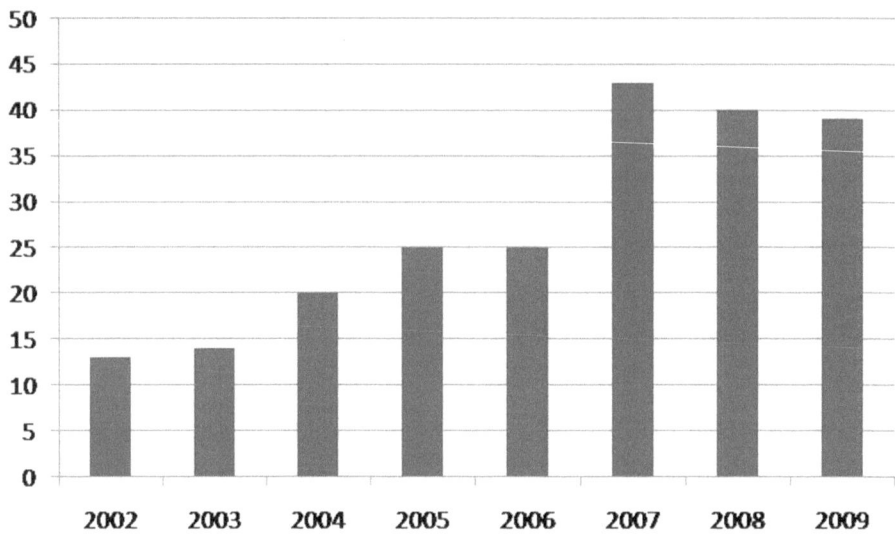

Figure 61. Annual activations of the International Charter for Space in Major Disasters.

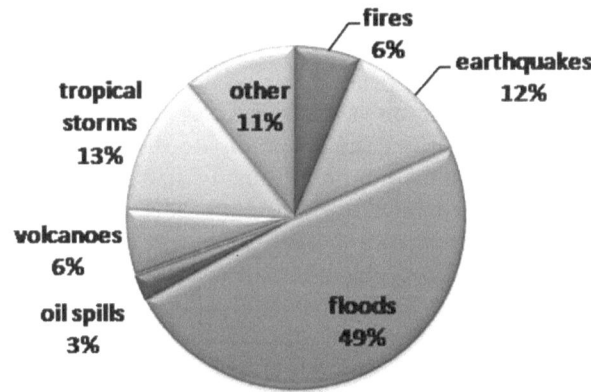

Figure 62. Reasons for activation of the Charter.

As shown in Figure 62, floods are the most frequent reason for which a member calls on the Charter. No doubt, this reflects the fact that floods can occur anywhere in the world and at any time, whereas earthquakes, for example, are infrequent and restricted to the earth's tectonically active zones. What Figure 62 fails to show is the relative severity of the individual disasters, so the relatively small number of earthquake incidents may well equal or exceed flooding as a cause of human misery and death. The 11% lumped into the "Other" category in Figure 62 includes tsunamis, landslides and accidents. Accidents that resulted in Charter activation included a shipping collision in the English Channel (February 2006), 100 ships caught in ice off Newfoundland (April 2007) and a massive explosion on a train in North Korea (April 2004).

Sir Martin and Surrey Satellite's success was not a foregone conclusion. When Queen Elizabeth II formally opened the new Surrey Satellite facilities in 1998, she remarked to Britain's Science Minister who was accompanying her that she hoped the government was helping Sir Martin in his endeavors. The irony was that the British Government had done nothing to help Surrey Satellite up to that point. Britain had invested large sums of money in earth observation satellites primarily through the European Space Agency (ESA), but these investments had focused more on radar imaging that was of interest to the military and involved large, heavy and expensive satellites – the Envisat satellite that carried an imaging radar developed in Britain cost over $1 billion, while one of Sir Martin's satellites cost less than 1% of that.

Perhaps stung by the Queen's throw-away remark, the British Government now directed some funding Sir Martin's way. Following the launch of their 17th and 18th satellites, one of which was a miniature demonstrator of the DMC concept, SSTL received a total of $15 million from the UK Government in 2000 to begin three satellite programs. One of the three was the British satellite in the proposed DMC constellation. This commitment of about $7 million by the British Government to DMC was relatively small compared to Britain's ESA initiatives, but was enough to persuade the DMC international partners to join. Algeria, Turkey, Nigeria and China each signed up for SSTL to build them a DMC satellite.

Another reason for Sir Martin's successful track record was that SSTL could build satellites not only cheaply and reliably, but quickly. Thus, it was less than 2 years later that the first DMC satellite was launched in 2002 for Algeria. The secret of SSTL's technical and engineering success is to use commercial electronic components wherever possible. This contrasts with the approach demanded by NASA, Department of Defense, ESA and other government agencies for specially developed "space-qualified" components to be used. The idea of "space-qualified" is that radiation from the sun is intense outside the earth's atmosphere and will destroy ordinary electronics. The vacuum of outer space also causes failure in many everyday components and mechanisms, so, again, NASA and other agencies require specially developed items to be used. What Sir Martin realized is that this is not a black-and-white situation – not all everyday components will fail, so by testing them, you can select those that will best resist radiation. Furthermore, Sir Martin's satellites all orbited below 1,000-km altitude and it is above that altitude that the radiation starts to get really intense – the Van Allen radiation belts discovered by America's Explorer-1 satellite in 1958.[2] Space-qualified electronic components are not only expensive (perhaps 100 times the price of commercial equivalents), but also are inevitably older and thus less powerful – it takes a few years to develop radiation-resistant versions of commercial components and in that time, faster and cheaper commercial ones will have appeared on the market.

---

[2] Instruments on the earlier Soviet Sputnik satellites had given a continuous maximum reading that the Soviets interpreted as a failure of the instruments, whereas Van Allen realised that what his instruments measured on Explorer-1 was a true radiation reading.

Figure 63. SSTL CEO Sir Martin Sweeting (right) and ESA Director Claudio Mastracci sign the Giove-A contract (with ESA officials in the background), 11th July 2003. Credit: ESA.

SSTL's ability to build fast, reliably and cheaply was demonstrated in a head-to-head race against the big two European satellite companies in 2003–2005. The European Space Agency purchased two prototype navigation satellites, Giove-A and Giove-B (Figures 63 and 64), as the beginning of Europe's version of the American military GPS system. Giove-A was built by SSTL for $34 million, was ready on schedule 30 months later and duly launched in December 2005. Giove-B was built by a consortium of Astrium and Thales Alenia Space, the two giants of the European space industry, for about $120 million but wasn't delivered until 2008, more than 2 years late. SSTL's Giove-A was guaranteed to only last for 2¼ years, but at the time of writing (2010) was still going strong. The much more expensive Giove-B is also working well, although it failed a few times in the first year of life. Both satellites have pretty much the same capability – Giove-B is more powerful in some respects, Giove-A in others. SSTL's successful delivery of the sophisticated Giove-A confirmed beyond any doubt its credentials as a mainstream satellite builder.

Sir Martin's DMC has been a success because by the start of the 21st century, commercial electronics were so powerful that a small satellite could achieve what would have required a giant satellite 10 or 15 years earlier. We see this improvement in electronics in our daily lives with the ever improving features in cameras, cell phones, television, laptop computers and the various hand-held devices for entertainment and information that didn't even exist 10 years ago. All of these products benefit from what is generally known as "Moore's Law" (Figure 65). This

Figure 64. Olivier Colaïtis, representing the Astrium and Thales Alenia Space consortium (left), and a slightly apprehensive-looking ESA Director Claudio Mastracci sign the Giove-B contract (with ESA official Joachim Schaper in the background), 11th July 2003. Credit: ESA.

"Law" is named after one of the founders of the Intel Corporation, Graham Moore, who said in 1965 that electronic devices would double in performance every 2 years or so.[3] That doubling continues today, despite many forecasts that it must stop. The "Law" applies not only to the speed of computers, but also to the size of memory and the number of pixels in a digital camera. All three (computer power, memory size and number of pixels) help small satellites do what could only previously be done by large ones.

Moore's Law allows engineers either to do more with the same size or do the same in a smaller size. The military tend to follow the first path – figure out the maximum-sized satellite they can get into orbit then get the best possible pictures from that machine. Commercial satellite companies tend to have smaller budgets than the military and thus aim to use smaller size to reduce costs. Cell phones, digital cameras and other domestic products tend to be a bit of both – using more powerful electronics but reducing size a bit, too.

Satellite images are not much use on their own. What aid workers need are up-to-date maps and satellite images can be used to produce such maps. The maps will show roads, names of towns and villages, built-up areas, rivers and lakes, railway

---

[3] More precisely, the number of transistors on a computer chip will double.

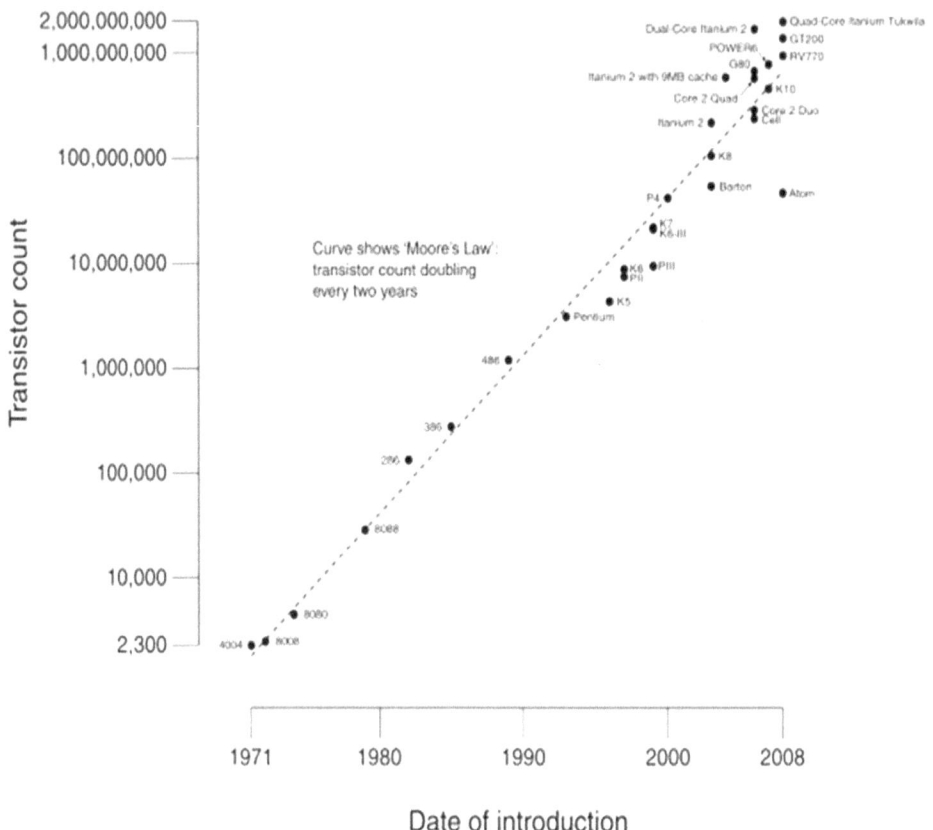

Figure 65. Moore's Law: inexorable improvement in performance of digital electronics.

lines and so on. The satellite images can be used to spot flooding, washed out roads, new housing or encampments. One straightforward way to do all this is to compare a satellite image from before the disaster with one from after. Changes between the two images indicate what needs to be changed on a map.

The UN has set up a unit to undertake just this disaster mapping function – the UNOSAT group based in Geneva, Switzerland. They respond to requests from other UN Agencies such as the UNHCR refugee and OCHA humanitarian agencies and have done so more than 100 times in the first 4 years since it began offering its "humanitarian rapid mapping service" in 2003.

While UNOSAT is a service for UN agencies, GDACS is a related UN service for any organization or person. The GDACS website (*www.gdacs.org/*) provides information on disasters around the world often within minutes of them occurring. Taking information from national and international agencies that monitor storms, earthquakes and volcanoes, GDACS is a "one-stop shop" for anyone interested in being alerted when a disaster occurs – you can sign up for an email or text message notification when a disaster occurs of the type you have expressed an interest in.

UN-sponsored initiatives such as the International Charter and UNOSAT are aimed at the formally organized relief agencies. Countries, regions, communities or individuals affected by a disaster may find it more difficult to get up-to-the-minute information. The media have a legitimate need to get reliable and current information, as do individuals in countries not directly affected who may have relatives or interests in the affected area. The UN-sponsored agencies under-standably resist getting drawn into providing information for anyone and everyone. They know that when a disaster happens, it will take all their available resources just to deal with those directly affected, so they are reluctant to divert people or resources for any other purpose. Europe's Respond program attempts to fill the gap. Although only funded for a 5-year period initially, Respond makes maps of disaster areas updated with satellite imagery available via the web.[4]

Almost exactly a year after the tsunami disaster, the DMC and Respond organizations were called upon again to help out. Yet another New Year was spent working 15-hour days as the teams provided rapid response images and maps for agencies attempting to help the victims of a major earthquake in northern Pakistan.

It's one thing to get images and maps immediately after a disaster; the next challenge is to get the information to the people who need it – usually in the field. Emergency agencies rely heavily on communications satellites for this. Typically, their work is in an area without modern infrastructure or where that infrastructure has been damaged. In those circumstances, satellites are the communications bearer of choice.

In fact, a specialist aid agency has been created to provide communications satellite services to crisis areas. Télécoms Sans Frontières (TSF), like its more famous near namesake Médecins Sans Frontières, is based in France. Founded in 1998, it has helped survivors and agencies in more than 60 countries. In 2006, the UN signed its first ever worldwide agreement with a non-governmental organization (NGO) – a "First Responder" agreement with TSF. "There is an urgent need for food, water, shelter, protection and medical help in emergencies," said Ann M. Veneman, Executive Director of UNICEF. "None of these things are possible without quick and reliable communications. Rapid communications saves lives." The founders of TSF Jean-François Cazenave, its President, and Monique Lanne-Petit, its Director, noted that "Access to reliable communications in the first hours following a crisis strengthens the coordination of agencies that save lives and respond to the needs of victims".

Mr Cazenave recalls that when he used to work for conventional aid agencies in crisis zones, victims would seek him out and press a piece of paper into his hand with a telephone number and name written on it, asking him to call that number when he returned to civilization. He noticed the high priority that many victims gave to establishing contact with family members, to report that they were safe or to find out about other family members. This observation led him to found his organization

---

[4]   See *www.respond-int.org*.

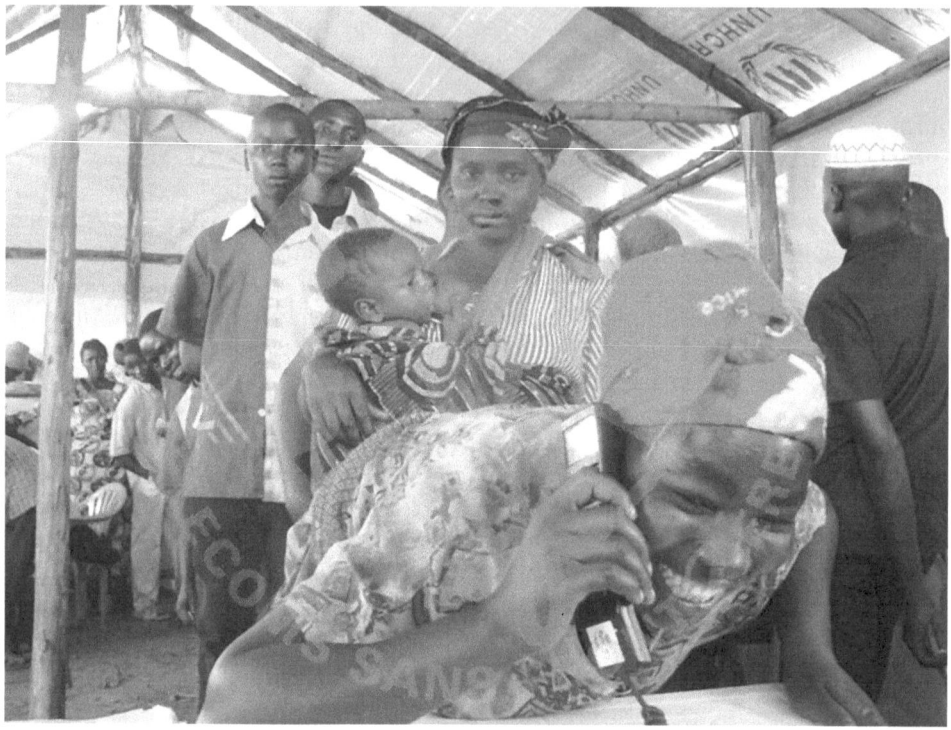

Figure 66. TSF users lining up, Matanda Camp, Uganda, December 4th 2008. Credit: TSF.

initially with a single Inmarsat satellite phone. Now, TSF operates with a dozen permanent staff and about 40 volunteers from three centers in France, Nicaragua and Thailand. TSF depends on the charity of organizations like Inmarsat, Eutelsat and Iridium to donate satellite time free – as you might expect, satphone charges are more expensive than a cell phone.

In November 2008, a TSF team was deployed to Uganda to help meet the needs of growing refugee populations fleeing the Democratic Republic of the Congo (DRC) (see Figure 66). The telecom centers benefited 10 humanitarian organizations and phone calls were provided for almost 2,000 families. The calls are often to give and receive family news, but requests for help and especially money are another regular motivation. In February 2009, a team of emergency TSF specialists deployed a Telecom Center in Dungu inside the DRC itself.

Perhaps the longest-running disaster demanding satellite images has been the man-made horror of Darfur in western Sudan. Situated on the southeastern edge of the Saharan Desert, Darfur, at the best of times, supports only subsistence farming and recent times have been anything but the best. A mix of tribal, religious and economic tensions have led to local people being terrorized by armed militias supported by the government. Refugee camps have sprung up to accommodate the tens of thousands of displaced families and international relief agencies have tried to bring food, shelter and water to the camps.

The very high-resolution imagery from the latest generation of satellites such as GeoEye, Ikonos, Quickbird and Worldview-1 and -2 is helping humanitarian agencies to monitor refugees and displaced communities. Frequently, these communities are in areas that are too dangerous for aid agencies or the media to access, so satellite imagery is often the only way to get objective information. For example, the Zimbabwe Government disputed UN reports that large numbers of people were displaced during Operation Murambatsvina (Restore Order) in 2005. Analysis of Ikonos imagery before and after the event showed clearly the leveling of a complete village at Porta Farm and enabled the number of displaced people to be estimated – about 4,300 at Porta Farm alone. These figures from just one small area of the country gave credence to the UN estimate of 700,000 displaced people nationwide.[5]

Satellites are effective in determining the extent of flooding and Hurricane Katrina provided a tragic opportunity to show this function in action. A number of satellite agencies produced maps from satellite images immediately after the disaster. The usual weakness of satellites is that where there is flooding, there is also likely to be clouds, making satellites like the Disaster Monitoring Constellation useless. The best such satellites can do is to wait for the storms to pass and the clouds to clear, and then take images of the flooded areas.

As discussed further in Chapter 8, a new generation of satellite is immune to the cloud cover problem because it uses cloud-piercing radar to build up images. Imaging radar satellites have been so expensive until recently that only the military and the occasional space agency could afford them.

Since 2006, Germany, Italy and Israel have launched satellites with imaging radars that don't cost the earth. Instead of the billion-dollars-per-satellite price tag of the US Lacrosse military and Europe's Envisat scientific satellites, these new satellites cost an order of magnitude less. Germany paid $400 million for the five-satellite SAR-Lupe constellation, Italy's four somewhat larger COSMO/SkyMed satellites cost $1,400 million, Germany's single TerraSAR-X cost about $230 million, its successor Tandem-X about $200 million, and Israel's TecSAR probably cost about $150 million. Other countries have joined or are planning to join this community, including Argentina, India, South Korea and Spain. The radar imaging satellites mentioned above can take pictures that resolve objects of 50 cm to 1 m in size. However, they often deliberately take less detailed images in order to observe a wider field of view – in the same way as you would decide whether to use the zoom feature on your personal camera or not: zooming in for detail, zooming out for a wide scene.

A normal earth observation satellite takes a photo of the scene below it just as you take a photo with your camera. If it's nighttime below, the photo comes out black and even in daytime, cloud appears as cloud. A radar imaging satellite works quite

---

[5]   Lavers (2010).

differently – it transmits a radio signal and detects the echo of that signal as it bounces off each point on the ground below. The ground is perhaps 600 miles below, so the echo is very, very weak. The satellite then assembles the echoes into the form of an image and sends that image to its control center on the ground. The radio waves are just as effective at night as during the day and are barely affected by rain, cloud or fog, so this type of imaging is able to provide pictures 24 h a day, whatever the weather.

Imaging radar is not perfect. It's quite hard to make out objects. A bright spot means that the surface is flat and pointing at the satellite, thus producing a strong echo – this might, for example, be a dark metal object that would appear black in a conventional photo. Likewise, a dark area in a radar image means that the surface gives a poor echo, but has nothing to do with how brightly colored it is – this might be because the surface is smooth and either flat (calm water, say) or pointing away from the satellite. The characteristics of imaging radar will be discussed further in Chapter 8.

The US military have had access to high-quality space imaging radar for nearly two decades and have no doubt worked out how to interpret these strange images over the years and civilians are now gradually beginning to do so. Radar images can pick up information that is a bit like color and again experience is needed to work out what it means. If more than one radio frequency is used by the radar or by two different radars, the echo will vary slightly and those differences are a form of color. Radio signals can have two different polarizations – a form of glint similar to the glint that Polaroid glasses remove from sunlight. The echo is slightly different for each polarization, so the differences can be considered a sort of color. Research groups around the world are busy trying to categorize how frequency and polarization can be used to make radar images more meaningful, but already it is possible to discern the maturity of some crops this way. Infoterra in France offers a service to farmers called FARMSTAR, which uses conventional satellite imagery to advise farmers about the progress of their crops. Imaging radar is used to fill in gaps in the information from visible images caused by cloud cover. The use of satellite images in farming was discussed in the previous chapter.

In Chapter 3, we saw how imaging radar has transformed the study of the polar regions. One of the techniques for monitoring polar ice is to spot changes between images taken at different times. This technique is also able to spot sinking or rising ground anywhere in the world and is being used to study volcanoes and earthquake-prone areas.

The LUSI[6] mud volcano in Sidoarjo, 700 km east of the capital Djakarta on Java in Indonesia, may have been a bizarre and tragic combination of man-made and natural catastrophes. An exploratory well was being drilled to look for natural gas in May 2006 and one theory is that it unintentionally released a rapid and unstoppable eruption of hot toxic mud. Another theory is that an earthquake 300 km to the west

---

[6]   LUSI = Lumpur (mud) Sidoarjo.

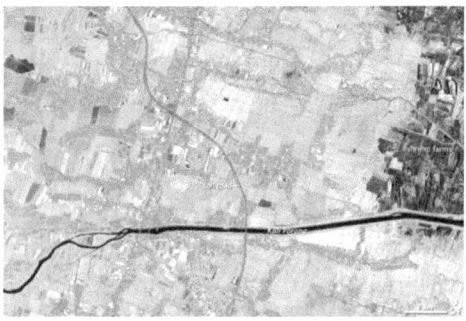 

28 August 2004   20 October 2009

Figure 67. Before and after images by NASA's Terra satellite of the area of the Sidoarjo mud volcano. Courtesy NASA Earth Observatory.

2 days earlier triggered the eruption. Thirteen people have been killed and 30,000 have been displaced from their homes. An area of 6½ sq km has been inundated despite the construction of levees to try to retain the flow and the eruption at a rate of 100,000 cubic m per day may continue for decades. Figure 67 shows the area in 2004 before the eruption and in 2009 with the man-made levees evident by their straight sides. In addition to the threat of inundation by the mud, the danger of subsidence is a concern – the removal of vast quantities of mud from below ground may cause the surface to collapse, especially weighed down by that same mud. The imaging radar on Japan's Daichi satellite is providing measurements of the subsidence. Images taken at 6-month intervals enable very accurate measurements of the difference in surface height in the intervening period. In the first 6 months, the ground subsided by more than a meter and it is continuing to subside at a lesser rate.[7] Daichi's radar is particularly suitable for this analysis because the wavelength of the radio waves of its radar are longer at 24 cm than those of most other space-based imaging radars.[8] The relatively short wavelength of other space imaging radars makes them too sensitive to the constant movement of foliage, vegetation and water ripples.

Comparing radar images taken at different dates allows very accurate measurements to be made of the changes in the surface between the two dates – at least in principle. We saw in the previous chapter that the technique is useful for detecting ground subsidence that might impact construction projects. Scientists are also using the technique to study the effects of earthquakes and volcanic eruptions, and hopefully to detect impending quakes and eruptions by spotting small movements of the ground in advance.

---

[7] Thomas *et al.* (2010).
[8] See Table 5 in Chapter 8.

WATER

Fresh water is an increasingly scarce resource in many parts of the world. The causes include: a rising population and a richer one (each person wanting more water); industrial pollutant, fertilizer and waste run-off; saltwater contaminating coastal aquifers as groundwater is depleted; and increased aridity in some regions due to climate change.

The issue is partly political because, as the saying goes, "water usually runs downhill, but it always runs uphill to money", implying that those with money get water while the poor go without. This leads to a distinction between *physical* water scarcity, where demand outstrips the available water supply, and *economic* water scarcity, where sufficient water exists but technical, political or economic weakness prevents it being tapped. Regions with physical scarcity today include much of the US southwest and adjacent Mexico, North Africa, southwest Asia, southern India, Pakistan, central Asia, northern China and southeast Australia.

One result is that the world's big three grain producers (China, India and the USA) are all experiencing falling water tables – China produces the world's largest wheat crop and it has fallen by 8% in the decade since 1997. Water tables are usually replenished by rain and rivers, but so-called "fossil" aquifers that consist of ancient water are not. In northern China, the immediate water table has been depleted and wells are now being drilled into the deep fossil aquifer. The World Bank foresees catastrophic consequences unless water usage and supply are brought into balance there.

Since the end of World War II, farmers in the mid-west of the USA have drawn from the fossil Ogallala Aquifer to irrigate America's breadbasket, but the rapid decline in the water table shows that this is not sustainable. The aquifer is being replenished on a sustainable basis in the north of the eight-state region it underlies. But its southerly portion has declined by as much as 50 m, especially in the Texas panhandle, west Kansas and east Colorado. Overall, the Ogallala Aquifer is being depleted today by an amount equivalent to 18 Colorado Rivers. The states are gradually taking action to address the situation – Kansas, for example, now restricts pumping to a sustainable amount and requires meters to be fitted to wells. Emptying the underground aquifer threatens the farming communities in the US mid-west. It also threatens wildlife. About half the shorebirds that migrate up the Great Plains stop at the Cheyenne Bottoms wetland in Kansas, which has been badly hit by depletion of the aquifer.[9]

Satellites monitor changes in lakes and rivers across the world simply by taking images on a regular and comprehensive basis. Already introduced in Chapter 4, a new method for satellites to monitor large-scale underground water aquifers has recently been demonstrated. NASA's GRACE mission measures the pull of the earth's gravity with unprecedented accuracy – so much so that it can detect small

---

[9] Rogers (2008), Brown (2009), Little (2009), and Weidensaul (1999) p. 313.

changes in gravity due to removal of water from the aquifers. GRACE has shown that in the 6 years to October 2008, the groundwater in Northern India was depleted at a rate of 4 cm per year or by a total volume of 18 cubic km per year. The World Bank reckons that 15% of India's food is produced using groundwater. The 114 million residents of the three-state region that includes the capital, New Delhi, risk depleting the underground aquifers completely due to extracting too much water primarily for agriculture. Rainfall during the period was close to the norm of 50 cm (20 inches) per year in what are essentially semi-arid regions. GRACE works by measuring the earth's gravity at a level of detail that detects features a few hundred km in size. It consists of two satellites orbiting one behind the other, separated by 220 km. The distance between the two is measured very accurately and changes in that distance indicate the variation in the earth's gravity below. Northern India is about 1,000 km across, allowing the broad-scale features of its gravity to be measured by GRACE. A decrease in gravity was detected in northwest India over the 6-year period, and the analysis attributed it to the sinking water table. A gain in groundwater further south was attributed to heavier than normal rainfall there during the period (see Figure 68).[10]

The same technique indicates a steady drop in the underground aquifers in California since GRACE measurements started in 2003 up to 2009 (the most recent data available) (see Figure 69). The drop is especially pronounced in Central Valley, which produces 8% (by value) of the food in the USA.[11]

There is no need for sophisticated gravity measurements to see the loss of fresh water in some other parts of the world. Satellite images are often sufficient to dramatize and demonstrate the loss of fresh water resources (see, e.g., Figure 70). This particular case is Lake Chad, whose shores are shared between four countries: Chad, Nigeria, Niger and Cameroon in sub-Saharan Africa. Overexploitation of the rivers that flow into it coupled with a drop in annual rainfall has reduced the size of the lake by 95% in the past 30 years. From being Africa's third largest lake, and the only one of significance in this part of Africa, the lake has now all but disappeared. The catastrophic impact on local societies and economies is not hard to imagine. As illustrated in Figure 71, Nigeria uses its DMC satellite, called Nigeriasat-1, to monitor Lake Chad in detail. The area shown in Figure 71 corresponds to the lower right-hand end of the Lake in Figure 70.

The bad environmental practice of the Soviet Union has become all too clear to see since the end of the Cold War. Central planning as practiced by the Soviets too often blinded the authorities to the environmental impact of their policies. The toxic radioactivity around the Chernobyl nuclear power plant in the Ukraine is well known. Less well known but even more toxically radioactive is the region around Chelyabinsk, east of the Ural Mountains, where Soviet nuclear weapons were manufactured. It is said to be "the most contaminated spot on the planet", with the

---

[10]   Rodell *et al.* (2009).
[11]   Bethune *et al.* (2009).

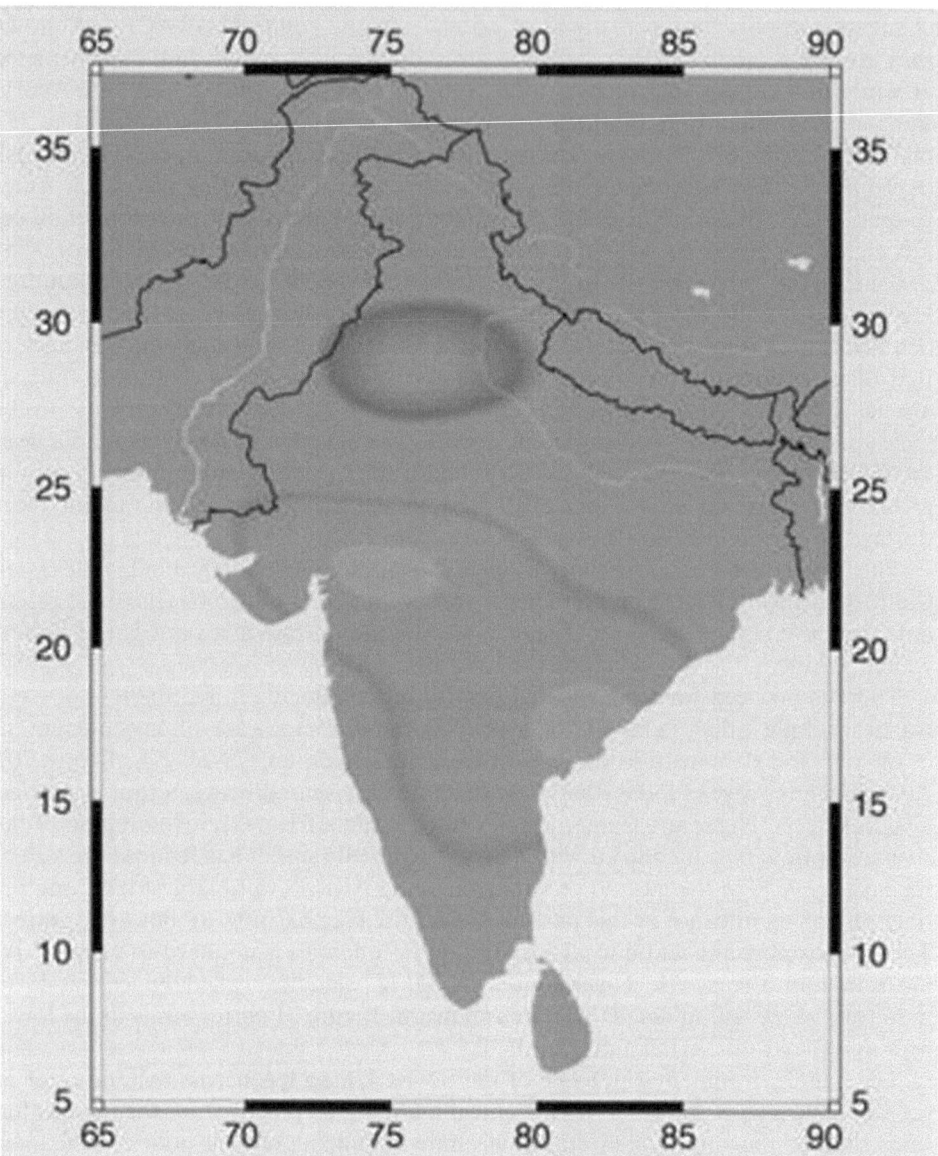

Figure 68. Area of groundwater loss in northern India (dark ellipse) and of groundwater gain in central and southern India (fainter contours) 2002–2008. Credit: I Velicogna/UC Irvine.

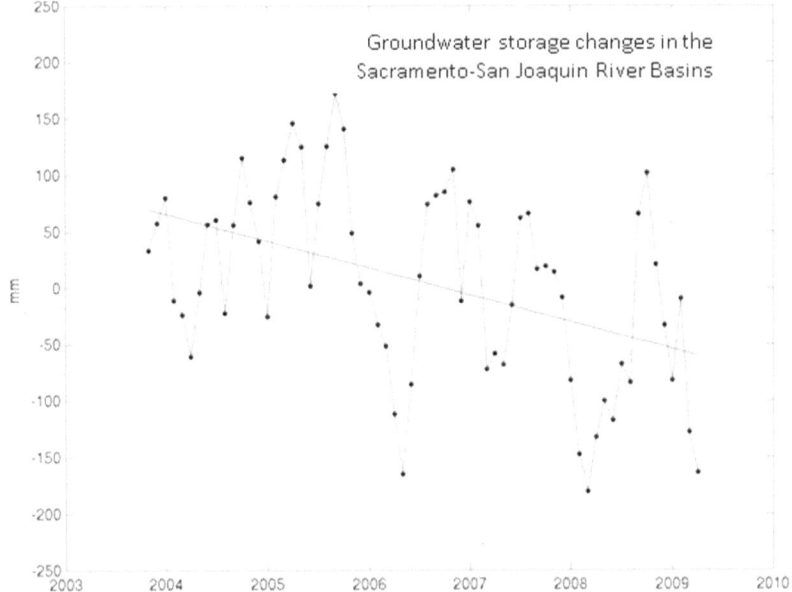

Figure 69. Depletion in Californian groundwater, October 2003 to March 2009. Courtesy NASA/JPL-Caltech.

years of casual waste disposal culminating in an explosion in a nuclear waste tank in 1957 resulting in a huge release of radioactive particles.

Figure 72 shows another example of Soviet environmental mismanagement. The Aral Sea used to be the third or fourth largest lake in the world.[12] The 30-year sequence of satellite images in Figure 72 shows the steady reduction in its size to the point at which what little water remains has become so saline as to be poisonous for most purposes. A whole shoreline community has been obliterated. The cause of the decline is the tapping for agricultural irrigation of the rivers that feed the Aral Sea.

Like Lake Chad, the Aral straddles an international border – that separating Kazakhstan and Uzbekistan. The inflowing river water primarily comes, or should come, from the Uzbekistan side, and the rivers also flow through Turkmenistan for over 200 miles. As in many parts of the world, and as illustrated starkly by these satellite images, agricultural progress often comes at the expense of someone else's fresh water resources.

---

[12]   Lake Victoria in Africa was about the same size and both laid claim to the "3rd largest" title at one time or another.

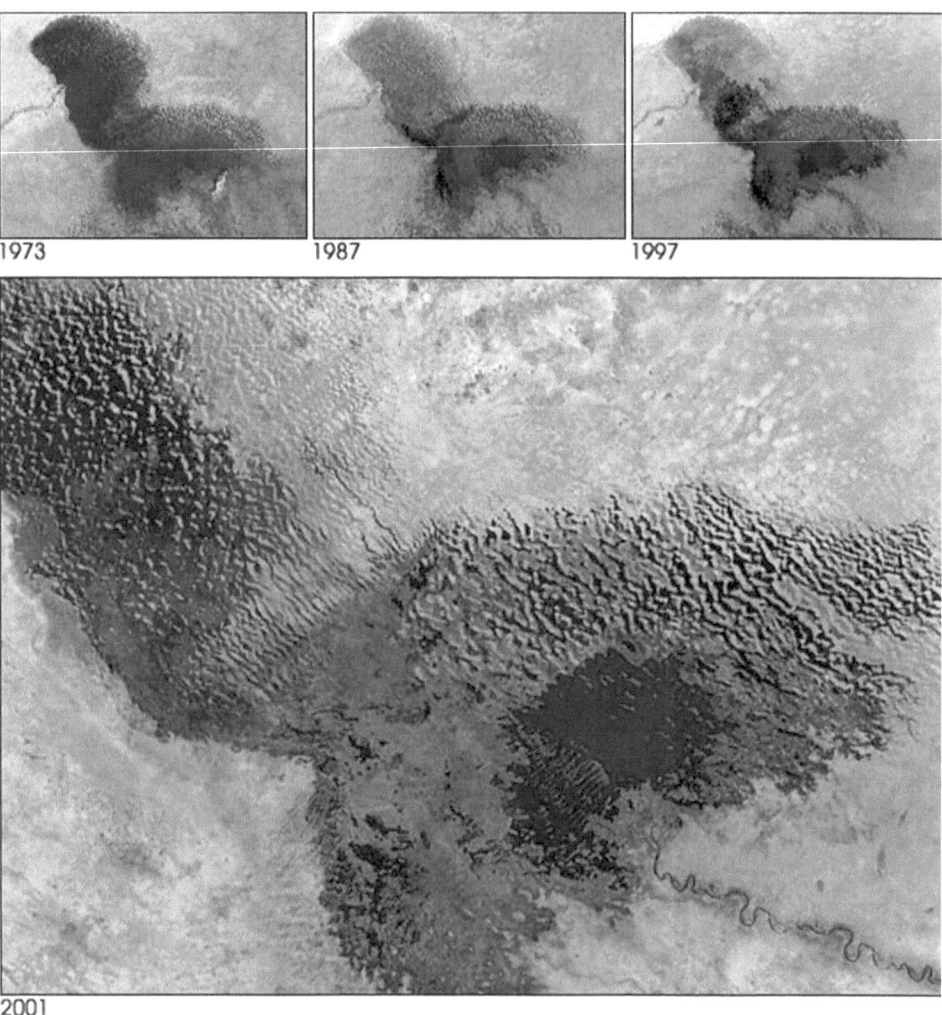

1973   1987   1997

2001

Figure 70. Lake Chad in 2001 and (insets at top) in earlier years. Credit: NASA GSFC Scientific Visualization Studio and Landsat 7 Project Science Office.

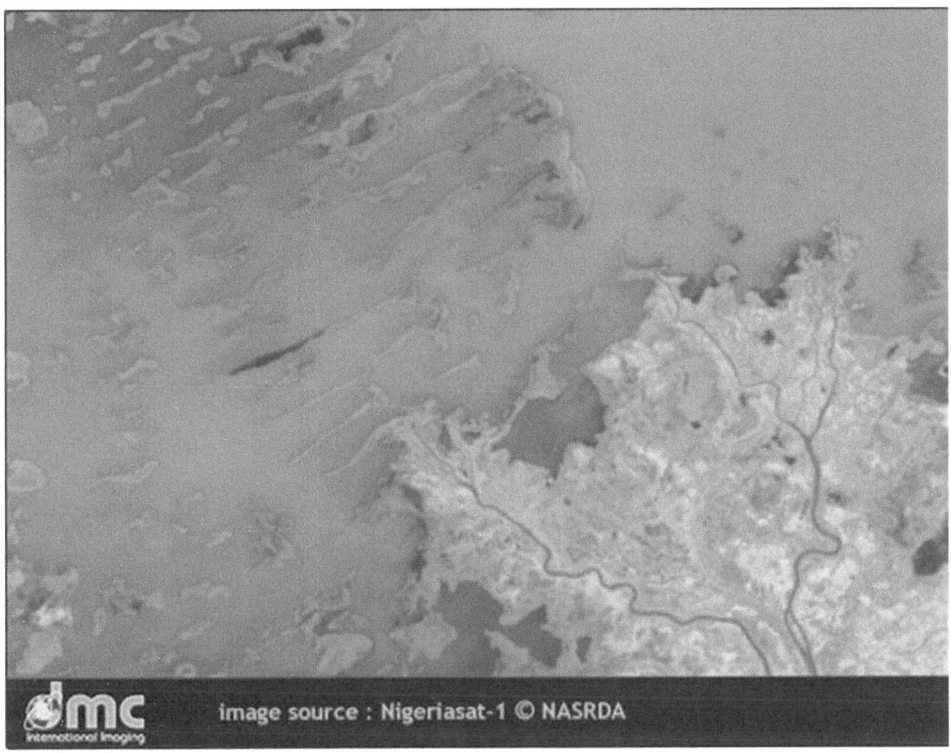

Figure 71. Lake Chad monitored by Nigeriasat-1, June 2005. Credit: DMC International Imaging Ltd.

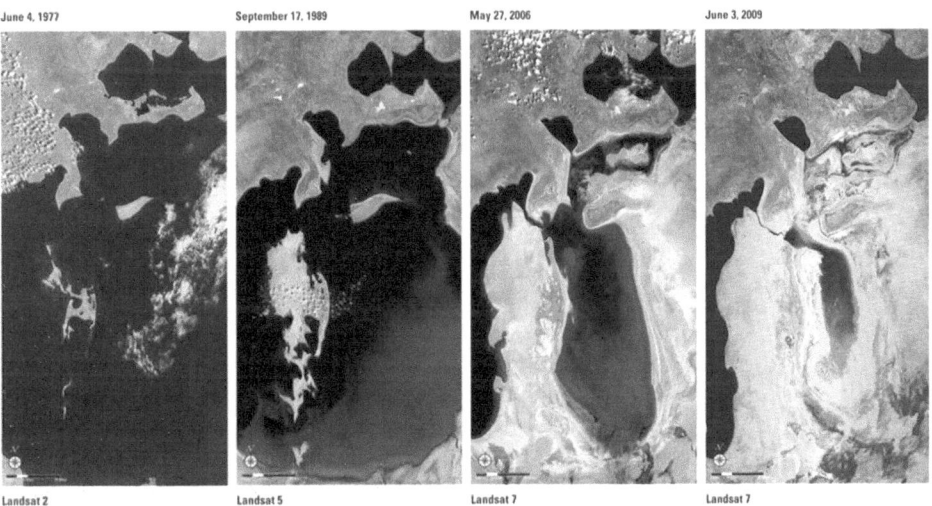

Figure 72. The disappearing Aral Sea (1977–2009). Credit: US Geological Survey.

6

Where am I? Where are they?

THE SURVEILLANCE SOCIETY

Satellite navigation (satnav) allows the world to know where you are – right? While traveling in a car (you wouldn't do this if you were the driver, of course), you call from your cell phone to get traffic information and the service tailors the response to where you are – because satnav has told them your location. You fire up the mapping service on your BlackBerry or iPhone and the service downloads the maps just for the area where you are. You use your iPhone to check where your friends are – so it's not just you that is being tracked. It seems as if those satellites in outer space are keeping track of where you are. They seem to be watching you.

The truth is more mundane. Your satnav device (cell phone, in-car satnav, iPhone, etc.) is watching the satellites, not the other way around. By doing so, your device can work out where you are (how it does this is explained below). Your device then tells the phone company where you are. Your friends' devices are doing the same, which is how you can find out where *they* are.

We unwittingly tell the world where we are many times each day. If a credit card is inserted in an electronic reader, the card issuer is notified. If you take money from a cash dispenser, the bank is notified. If you use your cell phone, the phone company knows which cell you are in, because your phone signal is routed through the nearest cell tower, each of which deals with phones in its local area or cell. There are more than 100,000 cell towers in the USA so that gives the phone company a good idea of where you are. If you drive your car past a speed camera or a toll booth, your license plate might be read by an automatic system and the information sent to police or other authorities. Perhaps the banks, the phone companies, the speed cops and all the other organizations that collect similar information are linked together and share this information. It's technically possible for every person to be tracked in this way, but are these government and commercial companies really so organized and competent?

I traveled by car to the annual conferences of the two main British political parties

P. Norris, *Watching Earth from Space,* Springer Praxis Books,
DOI 10.1007/978-1-4419-6938-5_6, © Springer Science+Business Media, LLC 2010

recently – one in the north of England, the other a week later 300 miles away in the south. As I approached the city in which the second conference was being held, I was pulled over by the police and asked the sort of questions police ask – name, address, purpose of trip, etc. Having established that I wasn't a security risk, the police explained that my car had been recorded at the previous week's party conference and that they were checking all cars that had done so. This minor episode in the battle against the terrorists illustrates the ability of the authorities to use identity and location information picked up from cameras – at least some of the time.

TV and Hollywood would have us believe that every time we pass a CCTV camera on the street, in a store, in the lobby of a building, at a train station, on a train or bus, etc., our face is analyzed to reveal our identity. This is fiction – for the moment. Yes, the technology exists to make this happen, but the quality of most of the CCTV cameras is woefully inadequate for the task. A blurred image is a blurred image, no matter how clever the software used to analyze it – "garbage in, garbage out", as the saying goes, meaning that if the picture being analyzed is poor, then the results of the analysis will be rubbish.

A natural extension of this technology is planned in the Netherlands. All cars will be required to carry a "black box" that logs their movements so that they can be charged for the use of the road infrastructure. The fees charged will reflect the degree of traffic congestion – more expensive at rush hours and in city centers, for example. Satellite navigation will be an essential part of this program. A large-scale test involving 60,000 cars is planned for 2011 before committing to the full nationwide system. Heavy trucks will be first to use the system, probably starting in 2012, with all road vehicles subject to it by about 2018. There are concerns about privacy and the abuse of the data by the public authorities. However, a recent case in Germany has tipped the balance of public opinion towards giving security agencies access to this sort of information. Police established that a serial killer was a truck driver and were sure they could identify him by analyzing the data from the road-charging system. Their request to access the data was turned down by the courts, even though it could have saved further deaths.[1]

The concern about satnav "watching us from space" is fuelled by the same all-knowing-state, Big Brother paranoia that makes some people cross the street to avoid a CCTV camera.

WHY WE NEED SATNAV

But satnav is increasingly a benign and even essential piece of technology – despite its use by the Mumbai terrorists as seen in Chapter 1. Until the 1990s, a 911 (or 112 or 999, depending on country) emergency call came from a fixed phone whose location was known to the emergency operator – once they identified the number,

---

[1]   Grzebellus (2010).

they could tell the location. Now, more than half of all emergency calls come from cell phones and the emergency operator has to ask the caller where they are calling from. Often, the caller is in a panic or even traumatized. If away from home, they may not know off-hand their precise location and may have to look around for a street sign or another person with the necessary information. If it takes longer than 6 min for the ambulance to reach a heart attack victim, their chances of survival diminish rapidly; thus, every second spent figuring out the location of the victim is precious.

Another problem of the cell phone age is that more than one person may phone in the emergency – this was much less likely to happen when phones were fixed. The emergency operators have to figure out whether two calls refer to the same emergency or different ones – in reality, there may be five or 10 calls to consider, not two. If the two emergencies are the same, then they don't want to dispatch another response vehicle. However, if the second call refers to a different emergency, then it is crucial that they *do* dispatch a second response vehicle. Even if the two calls are about the same emergency, the two callers may use different language to describe their location – in the open air, they may use nearby shops, landmarks, bus stops, etc. to explain their position, each using different points of reference. Perhaps one of the callers is in a car and makes the call having driven past the scene so the driver has to figure out not only where he is now, but where he was a minute or so ago.

As mentioned above, the phone operator knows which "cell" the caller is in, but that may tie things down only to within a mile or so. Worse still, if the caller is driving in a car, the cell identity may no longer be the same as that of the emergency.

The answer is to have a satnav device in every cell phone. Then, when an emergency call is made, the location as worked out by the satnav can be given immediately and automatically to the emergency operator. I have been at presentations by emergency medical personnel pleading with the phone authorities to mandate satnav in every cell phone. They know that it would save lives!

HOW SATNAV WORKS

If it's not watching you from space, how *does* the satnav tell you where you are? It's a bit like taking your bearings in the countryside. You scan around the horizon and spot a particular landmark way off to your left and another a bit ahead of you to the right. You look at a map and find the two landmarks, then put your finger on the map in between them so that they are in the same position relative to your finger as the actual landmarks are to you – one to the left and far away, the other to the right and a bit closer.

Your satnav device looks around the sky and picks up a signal from a GPS satellite to the left, another to the right. A message in the signal from each satellite announces its exact latitude, longitude and altitude, which is like telling you where it is on a map. Another part of the message announces the exact time it sent the message (and I mean *exact*, to the nanosecond), so when your satnav device receives the message, it checks its internal clock and knows how long it took for the message

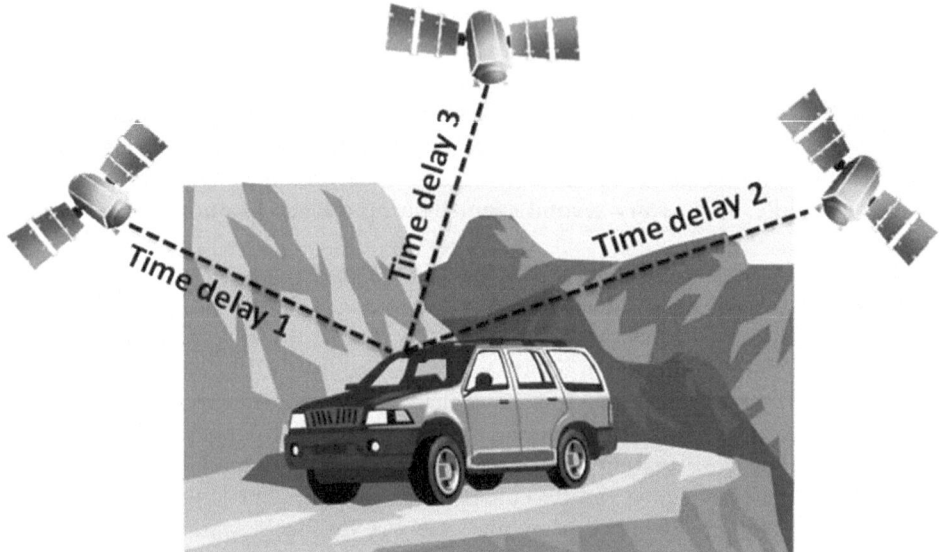

Figure 73. Three satellites give you your position.

to arrive, typically about a tenth of a second. Your device knows how fast radio signals travel and works out how far away the satellite was, typically 25,000–30,000 km. So, one satellite may be closer than another, just like the two landmarks you got your bearing from. With this information, the satnav device can figure out where on the map *you* are.

Figure 73 illustrates this but shows your satnav device tuning in to three satellites, not the two I described above. In practice, your device does indeed need to see three satellites because it needs a time check on its internal clock – the internal clock is not accurate enough for nanosecond work, so a third satellite allows your device to update itself. The clocks in the satellites are ultra accurate, but then they probably cost a million dollars or so each.

The three-satellite scheme illustrated in Figure 73 is fine if you are at sea level, but if you are in an airplane or in a mountainous region (as in the figure), then your satnav device needs to work out your altitude as well as your map position (latitude and longitude). This is accomplished by your device locking on to a *fourth* satellite, enabling it to calculate your latitude, longitude and altitude, and its own clock correction – four things to be calculated need four satellites. If the device can pick up more than four satellites, the calculation will be more accurate because it can take the average of several calculations. Typically, your satnav device can see between five and 10 satellites, unless you are in an area with high buildings or mountains that block its view of the sky.

VARIOUS KINDS OF SATNAV

The words "satnav" and "GPS" are often used inter-changeably, GPS being the US military Global Positioning System, which is by far the most widely used of the various satnav systems. Alternatives to GPS are listed in the panel below. GPS remains the gold standard for the sector and it is no surprise, therefore, that most of the alternatives follow the same principle as GPS whereby the user's device works out its own location based on information in the message transmitted by the satellites. It is understandable that what started out as a military system works this way, since you want soldiers, missiles, ships and other users to be able to work out their own location silently and autonomously. One implication of the information in the panel is that by 2015, there should be 100 or more navigation satellites offering GPS-like services. Provided your satnav device is compatible with them all, this huge fleet of satellites will ensure that you can get a navigation fix even in city centers or mountain valleys when most of the sky is blocked from view.

The world's main satellite navigation systems[2]

USA: Global Positioning System (GPS)
Thirty-one GPS satellites are in orbit – 24 is considered the minimum to provide continuous worldwide service. Each satellite circles the earth in about 12 h at an altitude of 20,000 km. Some transmissions are encrypted and on special frequencies intended for military users.

USA: Wide Area Augmentation System (WAAS)
The Federal Aviation Authority operates WAAS as an adjunct to GPS, alerting users in the USA (including Hawaii and the southern half of Alaska) within 6 s if any GPS satellite is not working properly and providing users with information to reduce the errors, especially those caused by electrical storms in the ionosphere.[3] Geostationary satellites are used to get the WAAS information to users rapidly and over a wide area.

Russia: Glonass
Glonass operates on a similar principle to GPS, with satellites about 19,000 km above the earth orbiting every 11¼ h. Twenty-two Glonass satellites are operational, with six more to be launched in 2010, giving the 24 plus spares needed for a continuous worldwide service.

Europe: Galileo
By 2014, Galileo will be similar to GPS, with 30 satellites 23,000 km above the earth orbiting every 14¼ h. Two prototype satellites are currently in orbit.

---

[2] As of July 2010.
[3] The ionosphere is the very tenuous part of the atmosphere from about 100 to 1,000 km above the earth.

Europe: EGNOS
EGNOS is the European version of WAAS (see above), enabling aircraft with WAAS devices to get a similar service over Europe.

China: Beidou/Compass
By late 2012, China is scheduled to have deployed four satellites in a GPS-like orbit (one already launched in 2007) plus 10 in geosynchronous orbits (three already in orbit). By 2020, the system should be global, with 27 satellites in GPS-like orbits.[4]

India: IRNSS
Seven satellites will be placed in geosynchronous orbit by 2014 to offer GPS-type signals over the South Asia region. The first launch is scheduled for late 2011.

India: GAGAN
GAGAN is the Indian version of WAAS (see above), enabling aircraft with WAAS devices to get a similar service over India.

Japan: QZSS
Three geosynchronous satellites will provide GPS-type signals plus some messaging services over the Japanese region. The first satellite is due to be launched in late 2010.

Japan: MSAS
MSAS is the Japanese version of WAAS (see above), enabling aircraft with WAAS devices to get a similar service over Japan.

International: Cospas-Sarsat
Twelve satellites currently carry the equipment to pick up signals from special Cospas-Sarsat emergency beacons. The system then computes the location of the emergency (some beacons contain a GPS receiver and give their own position) and passes the information to search-and-rescue authorities. Since 1982, it has provided distress alert information that has assisted in the rescue of 26,779 persons in 7,268 distress situations.

The philosophy of the Department of Defense (DoD) is to launch three GPS satellites each year, whether or not the existing ones have failed. Thus, if things are going well, there will be more than the requisite 24 GPS satellites in space, providing more targets for your satnav device to lock on to. The DoD is considering making 30 satellites the norm in order to increase the chances of seeing enough satellites in mountain valleys – a result of recent experience in the mountainous regions of Afghanistan. Various improvements have been introduced into the 2-ton GPS over the

---

[4]   Pirard (2010).

years, making it more accurate and resilient, and more are planned for the future. One consequence of the three-per-year launch rate is that it takes at least 8 years to have a full set of enhanced satellites in orbit. So, although the first of a new batch of 12 called GPS-IIF was launched in 2010, lighter than before at 1 ½ tons and each costing about $120 million (plus about the same again for the launcher), it is likely to be 2018 before its new features are available from all the satellites. The first of a major upgrade called GPS-III is due to be launched in 2014 costing about three times more, which will round off the enhancements of GPS-IIF and have additional ones of its own.[5]

The most mature of the alternatives to GPS is Russia's Glonass system. During the 1990s, as the Soviet Union fell apart and Russia sank into an economic crisis, Glonass was allowed to run down. With its economy now on a stronger footing (although staggering from the effects of the 2008–2009 world economic crisis), Russia is committed to making Glonass fully operational. It is on course for that to happen by the end of 2010, by which time a fleet of 27 or 28 of the 1 ½-ton satellites will be in orbit. Not only will the fleet of Glonass satellites be complete, but each satellite is much improved over its 1990s counterparts – they last longer before having to be replaced and the signals give more accurate results.

The first of a new generation of Glonass satellites with improved accuracy and a more sophisticated type of radio signal is due to be launched in 2010. This will be the first Glonass satellite not to house its electronics in an air-tight box and will be half the weight, at ¾ ton. Instead, the electronics will be open to the vacuum of space – as has been the norm for satellites of almost all other countries since the start of the space age. Back on the ground, the clunky and expensive Glonass devices hitherto available to consumers should soon be replaced by modern miniaturized affordable satnav devices like those for GPS. Prime Minister Vladimir Putin has recognized the value of satellite navigation as an enabler for economic growth as well as its propaganda value. He recently stated that "Glonass will cover the entire globe", that "it should be commercialised" and that it has a distinct advantage over Europe's Galileo system that is often touted as the best alternative to GPS. Alexander Gurko, who runs the company that operates Glonass, is taking this vision seriously. "We are actively establishing a joint venture in India to produce navigation equipment," he said. The Russo–Indian joint venture will produce receivers for military and civilian use that pick up both GPS and Glonass signals.[6]

Glonass uses the same principles as GPS, illustrated in Figure 73. A fleet of satellites 19,000 km out in space (GPS is 20,000 km out) transmits time signals that allow users to triangulate their positions. The radio signals are different from those of GPS so the two don't interfere with each other. Both GPS and Glonass have an additional feature in that they broadcast extra signals that can only be understood by military users. Then, in the event of a war, the ordinary signals are switched off or blocked and only the military signals remain.

---

[5]   Butler (2010b).
[6]   RIA Novosti (2010), Moskvitch (2010), Coordinates (2010), and Fyler (2010).

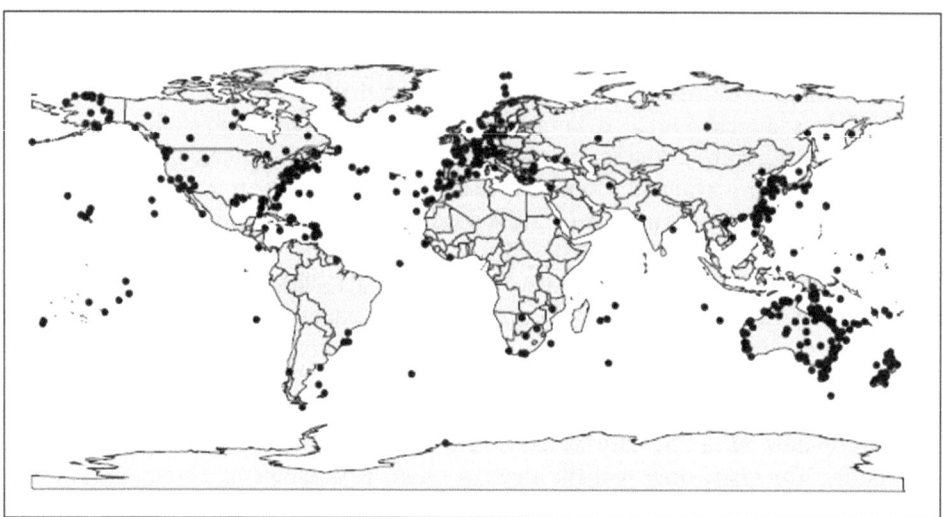

Figure 74. Emergency events for which Cospas-Sarsat data were used (2008).

The USA was unhappy when Europe decided to build a GPS look-alike system called Galileo. The European design was so similar to GPS that if the USA had blocked it in order to prevent the bad guys making use of it, GPS would have been blocked, too. Several years of negotiation resulted in Europe agreeing to make its design less sophisticated so that the USA can jam it (by blasting an area with high-power radio transmission, say) and GPS's military signals will still work. France was particularly unhappy about this compromise, since it had hoped to install the European system on the weapons it exports. France reckoned that it would gain an edge in export sales by claiming that the European satnav system could not be jammed by the USA. For the moment, that sales tactic is not viable – but some skeptical American commentators worry that France will figure out a way to subtly alter the European system to make it impervious to American jamming. An example of a conflict scenario in which this would matter is the 1980s war between Britain and Argentina over the Falkland Islands – a bunch of islands in the South Atlantic with a few thousand inhabitants. Argentina used French anti-ship missiles and warplanes, and inflicted heavy damage on UK forces with them. Had these missiles and planes been equipped with GPS satnav equipment, Britain might well have persuaded the USA to disable its civilian signals in the Argentina region, thus reducing the effectiveness of the weapons.

There is some concern that China's Beidou system will be uncomfortably similar to GPS and/or Europe's Galileo. As on many other aspects of trade between China and the West, this will no doubt be the subject of tough negotiations.

A total alternative to the GPS navigation technique is illustrated by the Cospas-Sarsat system. The key to understanding its purpose is the "Sar" at the start of the second half of its name, standing for Search And Rescue. The dozen or so satellites carrying Cospas-Sarsat equipment pick up distress calls from ships, planes, hikers

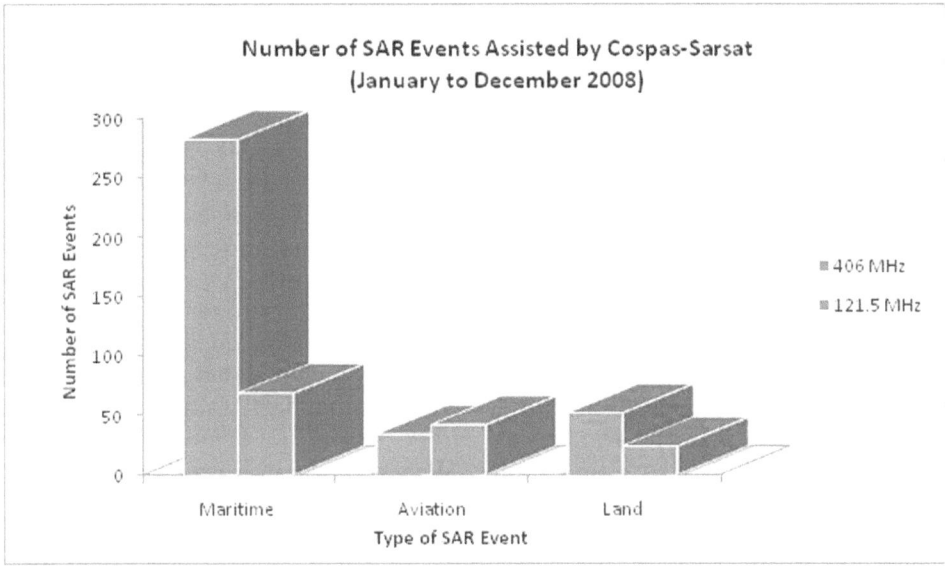

Figure 75. Cospas-Sarsat helping emergencies on land, at sea and in the air.[7]

and other travelers in remote areas and pass the information to the world's search-and-rescue authorities – 41 countries are signed up. A small office in Montreal, Canada, looks after day-to-day coordination of the system. The expensive part of the system is the equipment on the satellites and this is paid for on a voluntary basis by the countries that own the satellites. Each ship, plane or other user purchases the special beacon, which, in some circumstances, is mandatory to carry. Figure 74 shows the broad spread of the emergencies supported by Sarsat-Cospas and Figure 75 quantifies the preponderance of those that involve ships at sea.

Cospas-Sarsat works because users broadcast the fact that they need help. Originally, the user's beacon didn't have to tell the system its location; instead, the satellites recorded the signal for the 10 or so minutes while it passed overhead and the ground controllers could work out the location of the emergency from the changes in the signal. The technique uses what is called the Doppler shift in a radio signal as the transmitter approaches or leaves the receiver – just as the tone of a police siren rises from a bass to a baritone and a tenor as it approaches and descends back to a bass again as it departs.[8] Nowadays, most emergency beacons incorporate a GPS satnav device and are able to tell the Cospas-Sarsat system their location without the need for the Doppler-based calculations.

---

[7]   The left-hand bar in each category refers to a modern digital beacon, the right-hand bar to the older type of beacon.

[8]   The volume also rises and falls as the siren approaches and departs, but the tone or pitch or frequency is the Doppler phenomenon.

ACCURACY

The GPS concept of a silent user was devised to work without the user needing a very accurate (and thus expensive) clock. As mentioned earlier, your satnav device works out the error in its clock by gathering information from a third or fourth satellite. Another clever feature of GPS is that the signal was originally deliberately distorted so that it was less accurate for civilian users – about 50-m accuracy for you and me, versus 2 or 3-m for the military, who could access the undistorted signal. However, as explained by Professor Brad Parkinson (usually introduced as the "father of GPS"[9]), "as soon as the [first Gulf] war started, they decided to turn [the distortion] off since many of the soldiers had civilian GPS sets" (Figure 76).[10] What happened was that GPS was only partially operational, with about a dozen satellites in orbit when Saddam Hussain inconveniently invaded Kuwait in 1990. American and allied forces deployed quickly to the northern part of Saudi Arabia, preparing to invade Kuwait and expel Saddam's forces. Since GPS was not yet fully operational, the Army had not yet equipped all of the troops with GPS devices, but a vast featureless desert is exactly the sort of place in which GPS is needed. Back home, Mom and Pop stopped by Radio Shack and bought a cheap and cheerful GPS receiver and shipped it out to their sons and daughters in Saudi – by the tens of thousands! In theory, Saddam's forces could have done the same thing, but they didn't have such easy access to Radio Shack – and this was a decade before they could order them on Amazon. So, the grunts with military GPS sets knew their position to within a couple of meters, while those with civilian sets had to settle for 40–50-m accuracy. The Department of Defense recognized the benefit of giving all troops the full accuracy and made the civilian signal as good as the military one – at least for a while.

After that war, the DoD degraded the civilian signal once again, but this gave rise to another bizarre situation, as Brad Parkinson points out: "... the United States Coast Guard put together a system of marine beacons that was taking that [distortion] error out. So you had one group in the government putting the error in and another group of the government taking it right back out." Eventually, in 2000, the government announced it would stop degrading the civilian signal and give all users the military level of accuracy.

The military have certainly benefited from GPS – whereas 10 bombs were required on average to destroy a target in the first Gulf War (1991), that figure was reduced to one in the Iraq War (2003) because of GPS – with a consequential reduction in collateral damage. Brad Parkinson says that "GPS is a humanitarian weapon system! By that I mean you do not hit things you do not want to hit. The advantage of GPS based weapon delivery is that if the target is accurately located; you hit the target, you do not hit the nearby mosque, or church, or temple, or hospital."

---

[9]   Sanjai Kohli, co-founder of GPS chip manufacturer SiRF, is often called "the father of mass market GPS".

[10]   Parkinson (2010).

Figure 76. "Father of GPS" Brad Parkinson led the team that designed GPS. He argues that "GPS is a humanitarian weapon system". Credit: Stanford University.

Although he and his colleagues foresaw many of the ways in which GPS would be used back in the 1970s when they designed it, he admits that many of its current uses surprise him. As an example, he notes that "right now GPS Agriculture is a business worth more than 400 million dollars a year worldwide and growing", referring to technology that enables robotic farm tractors to pull an implement on a rough field with an accuracy of about 4–5 cm.

IN-CAR NAVIGATION

The first mass-market use of GPS was to provide driving directions in cars. This depends primarily on three technologies that became affordable at about the same time. The first is GPS, which tells the device its latitude and longitude. The second is a digital map stored in the device, which includes road names, traffic restrictions, speed limits and the like. The third is software to work out the best route between two locations. A color screen that can be controlled by touching it (to avoid having to use a keyboard) and a powerful miniaturized computer (so that it produces answers fast) are other general-purpose technologies needed.

Thus, for in-car navigation to work, you need all of the above. GPS alone tells you latitude and longitude, but without a map, that's not much use. Even with a map, it isn't much use because what you really want to know is "which turning do I take?" That requires a computerized map plus logic (software) to work out the fastest way from point A to point B. Thus, the combination of GPS, computerized map and route guidance software is the key to in-car navigation becoming so popular. If any of these contains errors, the experience can be unpleasant. People usually blame GPS, but in fact, it is much more likely that the error is in the digital map or the route calculation.

Even Google is not immune to the perils of GPS misadventure. A journalist using an early version of its mapping and navigation software in New York City was informed by the device that he was in London and was presented with options for reaching nearby London underground stations.[11]

As Brad Parkinson puts it, "GPS cannot replace common sense!"

PRIVACY AND SATNAV

One of the interesting uses of GPS is to enable very young and very old people to be tracked. Many cell phone operators offer a service whereby an authorized subscriber can be told where another subscriber is. The scenario used to sell this service is that of keeping track of a child or a senile adult. You equip the target person with a cell phone that is always switched on, perhaps sewn into their clothing or strapped to their leg, and you are then confident that you can find out where they are. Who could object to that?

Consenting adults can also use the service so that you can see where your friends are and vice versa. Cell phone operators have come up with lots of variations on the service, such as enabling you to identify which of your friends are within, say, 200 m of you. Koni, Vladimir Putin's black Labrador, has a Glonass-equipped tracking collar.

Of course, you must prevent unauthorized subscribers being able to see where your child is. One scheme is that you use the child's cell phone to tell the phone company which numbers may request information about it. But suppose a pedophile gets access to the child's phone for a few moments and authorizes his phone number. How do you prevent that? Another scheme is for the parent's phone to request that the child's number be tracked. Once again, the challenge is to prevent a pedophile requesting the child's number from his phone. Most phone companies choose this second approach in which the parent's phone controls the access and no other phone is able to request tracking privileges. The pedophile then needs access to the parent's phone for a few moments to insert his phone as an authorized tracker of the child. The phone companies deal with this by sending a message to the parent's phone a few hours after a new tracking authorization is received, asking the parent to confirm the request.

Well that's the theory. Some countries have introduced regulations or laws to formalize these arrangements. But consumer groups have found that the phone companies don't always live up to their promises, perhaps forgetting to seek confirmation about a new tracking request, for example.

---

[11]   Jaroslovsky (2010).

Cell phone companies have two ways to locate your cell phone. One way is by placing a GPS receiver in the handset. The other way is to triangulate your position from several nearby cell phone towers. Your cell phone is picked up by any nearby tower, but the phone company only uses the nearest one to handle the call. The signals picked up by other cell towers are ignored except if the phone company installs special equipment to perform triangulation. The advantages of GPS are that it is very accurate and costs the phone company nothing – you and the other five billion cell phone owners[12] pay a little bit extra (perhaps a dollar or two) to have a phone with GPS installed. The disadvantage of GPS is that it doesn't work indoors. Triangulation via cell towers costs the phone company the price of the triangulation equipment on every tower (thousands of dollars on each of the tens of thousands of towers) is accurate only where there are lots of towers such as in cities but has the advantage of working when the subscriber is indoors. The issue of authorizing people to track a child arises whichever location technology the network uses – GPS or cell tower triangulation.

GPS isn't going away – the cell phone services that exploit it are already a $1.6 billion business, and one analyst forecasts 780 million individual users by 2014.[13]

GPS *IS* WATCHING YOU

In one specialized way, GPS satellites *are* watching us. They carry special instruments to detect a nuclear explosion on earth. This unusual job is a legacy of the Cold War when the USA and Russia had satellites dedicated to detecting and locating nuclear explosions – thankfully, all the explosions were tests rather than as part of a war. The instruments on GPS detect only nuclear bomb tests that take place above ground, the last of which occurred in 1980. The underground nuclear tests that have continued intermittently since then are monitored by seismic detectors in laboratories around the world.

One of the instruments on the GPS satellites detects the bright flash of the fireball at the heart of an atom or hydrogen bomb explosion. As discussed in Chapter 8, that flash has a characteristic "twin peaks" shape that, as far as we know, is unique to a nuclear explosion. Other instruments detect the burst of X-rays and gamma rays (the most intense forms of light) and the electromagnetic pulse (EMP) emitted by the explosion.

The early Vela satellites that carried these instruments before GPS was around produced a scientific bonus by detecting intense bursts of gamma rays that came from distant stars or galaxies – these bursts had not been expected and are still the subject of research. Many of these astronomical gamma ray bursts are now thought to come from supernovas – the explosion of a giant star – almost all of which are

---

[12]   ITU (2010).
[13]   *www.iemarketresearch.com.*

very distant, way beyond our own galaxy. But others, especially those of very short duration, are still unexplained.

Astronomers have therefore benefited from these military satellites. Another group of scientists uses GPS to monitor the atmosphere. As the radio signal from GPS passes through the atmosphere to reach the ground, it is bent, delayed and distorted very slightly. The exact amount of change in the radio signal tells the scientists something about the amount of moisture in the air at that place and time. With thousands of GPS receivers all over the world, this information builds up into a useful pool of data for weather scientists.

7

Monitoring nuclear weapons[1]

America's first space program was to develop surveillance satellites to monitor the nuclear arsenals of the Soviet Union and its allies. Although the Cold War has ended, nuclear weapons remain a threat to world peace and surveillance satellites are still an important tool in monitoring them. A recent US policy statement on the subject stated that "The threat of nuclear war has become remote but the risk of nuclear attack has increased".[2] An all-out nuclear exchange between the USA and Russia looks very unlikely, but nuclear weapons are now held by nine countries, some in areas of the world where the threat of war is high.[3] Recent calculations suggest that a nuclear war between regional powers could cause a decade-long global winter, with a death toll of about a billion (mostly among those who live on marginal food supplies).[4]

This chapter outlines the steps taken by the international community to control the nuclear threat. In particular, it looks at how cameras in space can help to monitor the agreements on nuclear weapons signed up to by most countries.

THE COLD WAR: CORONA AND ZENIT-2

In 1954, 3 years before the Soviet Union launched the world's first artificial satellite, Sputnik, the USA began a secret program to develop a satellite that could

---

[1]   I am indebted for most of the post-Cold War material in this chapter to Robert Kelley, formerly a nuclear weapons expert from the USA who did weapons inspections for the International Atomic Energy Authority (IAEA) Action Team and the Department of Safeguards and now an Affiliated Senior Research Associate at the Stockholm International Peace Research Institute (SIPRI).

[2]   Gates (2010) p. 3.

[3]   The "nine nuclear countries" excludes otherwise non-nuclear NATO countries that "participate in nuclear planning and possess specially configured aircraft capable of delivering nuclear weapons"; Gates (2010) p. 32.

[4]   Robock and Toon (2010).

P. Norris, *Watching Earth from Space,* Springer Praxis Books,
DOI 10.1007/978-1-4419-6938-5_7, © Springer Science+Business Media, LLC 2010

photograph military installations behind the Iron Curtain. Supported strongly by President Eisenhower, the spy satellite program was a response to the culture of secrecy and deceit in the communist dictatorships. America had built the first atomic bombs in 1945, using them at Hiroshima and Nagasaki to help end World War II, and then the even more powerful hydrogen bomb in 1952. To the surprise of most in the West, the Soviets had exploded an atomic bomb just 4 years after the USA and a hydrogen bomb just 10 months after them.

The Soviet Union and other communist countries behind the Iron Curtain were police states. Information in the media was controlled by the state and comprised bland and blatant propaganda. Movement around the country required authorization and travel abroad was almost totally forbidden. Foreigners were not welcomed except as part of a state program, such as education, and could only travel in the company of a government "minder" and even then, only to state-approved destinations. The only publicly available uncensored maps of the Soviet Union pre-dated the 1917 communist revolution.

Military planners in the West obtained information on Soviet military forces from defectors, from radars and radio listening posts in countries bordering the Soviet Union and from occasional flights across Russia by America's high-flying U-2 spy plane. Information from the U-2 ceased in 1960 when the Soviet air defenses finally managed to shoot one down and capture the pilot. The Soviets launched Sputnik into orbit in October 1957 and thereby demonstrated all too vividly how they could place an object the size of a nuclear weapon anywhere on the planet. America's lack of knowledge about Soviet long-range rockets and nuclear weapons is illustrated by the official US estimates of Soviet military forces in the summer of 1960 (highly secret at the time, of course). Each of the US Armed Forces (Army, Navy and Air Force) made a separate estimate of Soviet Union long-range missiles, resulting in values ranging from 150 to 700. The CIA separately estimated 400. The Armed Forces estimates were self-seeking, with the Air Force postulating a large number of Soviet missiles that would require a comparable Air Force response, while the Army and Navy both offered a low estimate so as not to take funding away from their programs. The CIA estimate was explicitly in the middle and so gave no added confidence to the President. A year later, the first American spy satellites had returned good imagery of most of the Soviet Union, and the Soviets were found to have almost no long-range missiles; those they had were difficult to deploy in large numbers.[5] US Air Force "hawks" had been advocating an attack on the Soviet Union before their missile forces became unstoppable, but the spy satellite information undercut their arguments, thus sharply reducing the risk of nuclear conflict between the super-powers. The critics who had claimed that Sputnik pointed to a Soviet lead in long-range missiles were shown to be completely wrong and, in fact, the USA had a significant lead.

The US CORONA spy satellites recorded their images on a conventional "wet"

---

[5]   Norris (2007) p. 102.

film. When the film was used up, it was ejected in a protected capsule and returned to earth – being caught by an aircraft as it descended towards the ocean suspended below a parachute. Initially, each satellite ran out of film after a few days, although as the technology evolved, this became a few weeks. Throughout the 1960s, the USA launched CORONA satellites every month or so in order to keep abreast of Soviet and Chinese military developments. By the time a successor system called Big Bird was deployed in 1971, 120 CORONA satellites had been placed in orbit.

Figures 77 and 78 illustrate the type of information provided by CORONA and other similar US satellites. The image of the ICBM site at Yur'ya shows the configuration of missile silos, anti-aircraft defenses, approach roads, boundary fences, etc., all of which were repeated precisely for a specific type of ICBM and differed in various details from the sites of other types of ICBMs. Thus, this imagery enabled analysts to identify the number of missiles of each type deployed across the Soviet Union.

The image of the Chinese nuclear weapons test site at Lop Nur in Figure 78 shows the high tower in the center of the circular test area, indicating that a test was imminent. The prototype hydrogen bomb was placed on top of the tower before ignition.

The Soviets were a bit slower to recognize the opportunity to monitor its adversary's military assets by satellite. But they soon caught on, and from the first successful flight in 1962, 80 of the Soviet Zenit-2 spy satellites were launched during the 1960s of which 58 were completely successful.

In the Cuban missile crisis of 1962, the Soviets backed down publicly from their plan to deploy missiles in Cuba. As a consequence, the Soviet military determined not to be outgunned by the Americans again and began a rapid deployment of long-range missiles – initially in underground silos and then in submarines. The USA responded in kind and thus began an escalation of nuclear forces that resulted in both sides having more than 2,000 intercontinental nuclear weapons each by 1972 and over 3,000 each by 1979. The number of nuclear devices of all types was an order of magnitude greater (see Figure 79), being also deployed on short- and medium-range missiles, artillery shells, conventional bombs, etc. Both sides recognized that their nuclear arsenals were dangerously over-sized, but did not trust the other side enough to stop the escalation.

The CORONA and Zenit-2 satellites could detect and count long-range missile sites, missile-launching submarines and long-range bombers. This ability was critical in allowing the super-powers to eventually negotiate a halt to the seemingly inevitable increase in the number of nuclear weapons. The first Strategic Arms Limitation Treaty (SALT-I) 1972 slowed the increase and SALT-II 7 years later brought it to a halt. The super-powers didn't trust each other, but with their spy satellites, they were sufficiently confident of detecting infringements of SALT-I and SALT-II to overcome the mistrust.

Figure 77. Annotated CORONA images of SS-7 ICBM site at Yur'ya, June 1961 and June 1962. Credit: National Reconnaissance Office.

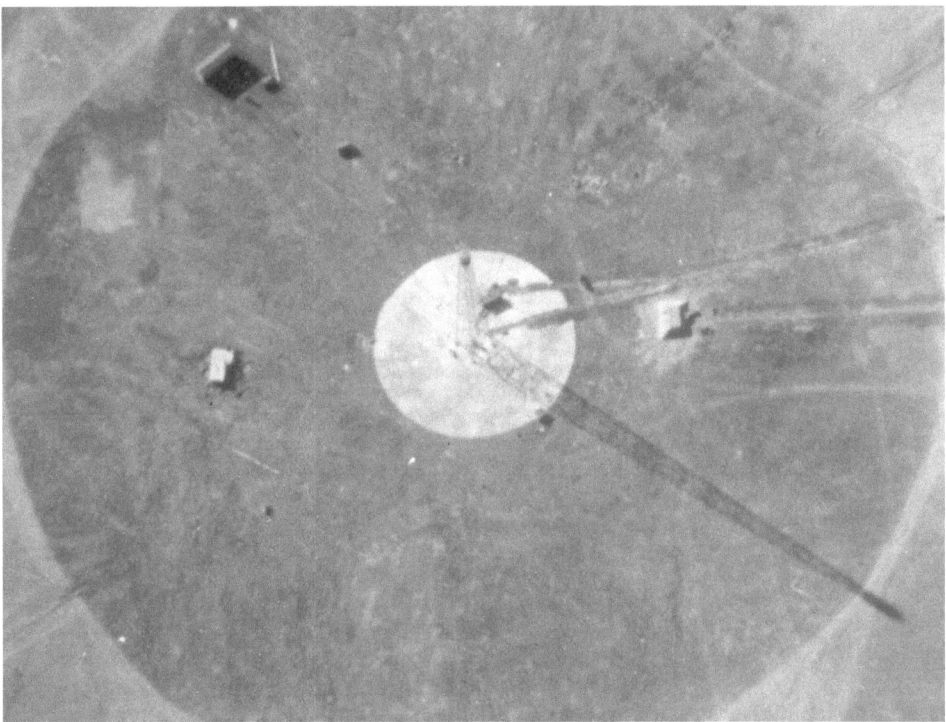

Figure 78. GAMBIT KH-7[6] image of China's nuclear test site at Lop Nur, 20 days before the December 28th 1966 test. Credit: National Security Archive.

THE NUCLEAR NON-PROLIFERATION TREATY

Even before the super-powers had found a way to negotiate SALT-I, they had found common ground in wanting to avoid other countries developing nuclear weapons. Pressure for some sort of agreement to limit the spread of nuclear weapons was widespread, spurred on by the realization that the radioactive fallout from any nuclear explosion would affect the whole world. The Nuclear Non-Proliferation Treaty (NPT) was negotiated, aimed at restricting the number of countries with nuclear weapons to those that possessed them in the mid 1960s. The cynics saw this Treaty as the major powers wanting to avoid other countries becoming a threat to them. It did indeed divide the countries of the world into two categories – the five nuclear powers (the Soviet Union, the USA, Britain, France and China) and the rest. It was clear why the five nuclear powers favored that division of the world and the beauty of the Non-Proliferation Treaty was that it provided an incentive for the

---

[6]   The GAMBIT KH-7 satellite was a US Air Force satellite with some features similar to the CIA-sponsored CORONA.

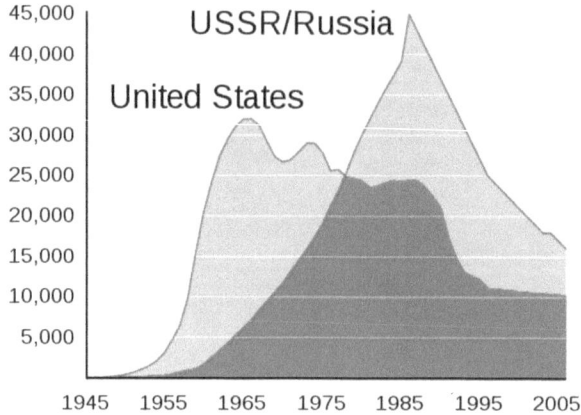

Figure 79. Nuclear weapon stockpiles of the super-powers. Credit: Wikipedia Commons.

other countries to accept their lot in the non-nuclear club. Note that although France and China possessed nuclear weapons, they didn't immediately sign the NPT, but did agree to abide by its provisions.

In return for being the only nuclear powers, the big five had to agree to help the other countries in various ways. Economic and technical assistance was promised to developing countries in the use of nuclear energy for peaceful purposes. The NPT in effect encouraged countries to build nuclear power stations, even though the nuclear fuel used in those power stations could become the starting point for developing nuclear weapons. Another important feature of the NPT was that the big five promised to pursue negotiations towards nuclear disarmament and, indeed, towards complete disarmament under international control.

The NPT opened for signature in 1968 and entered into force in 1970. It is one of the most widely accepted treaties in the history of arms control and was extended indefinitely in 1995. The NPT seemed to be standing the test of time until a huge clandestine nuclear weapons program was discovered in Iraq as a result of the Gulf War of 1991.

Initially, the NPT was focused on keeping track of nuclear-related material in countries with peaceful nuclear activities. The main principle behind preventing the spread of nuclear weapons was controlling the fissile materials used to make them. These materials, generally uranium enriched in the isotope ^{235}U (enriched uranium), and plutonium are the fissile materials that make nuclear bombs work. Only a few states had the resources to produce enriched uranium or plutonium and so controlling and accounting for materials were relatively straightforward tasks. By contrast, today, the International Atmoic Energy Agency (IAEA) considers that 72 countries have significant nuclear activities.[7]

---

[7]   IAEA Secretariat (2006) p. 9.

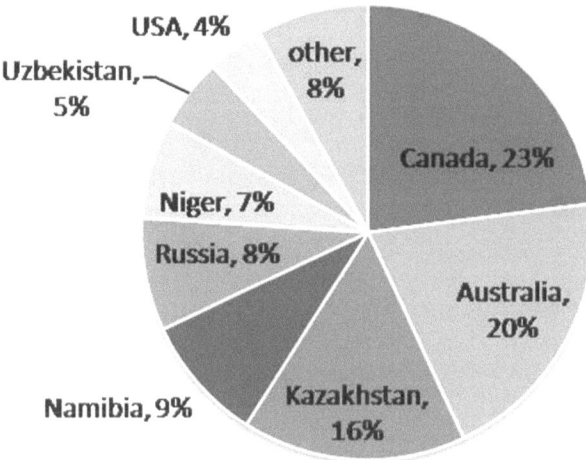

Figure 80. Distribution of the 42,500 tons of uranium mined in 2007.

FACILITIES TO PRODUCE NUCLEAR MATERIALS

Production of uranium

Plutonium does not exist in nature and must be made by subjecting uranium to intense neutron irradiation in a nuclear reactor. Hence, uranium mining is the first signature that analysts look for when searching for nuclear activities, whether the end product is plutonium or uranium. Uranium mining and milling are not visibly different from other mining activities, but detailed analysis of a mine and mill may lead to a high degree of certainty that uranium mining is taking place.

Currently, uranium is mined in 20 countries, with Iran being the latest entrant. Canada and Australia currently account for nearly half of the world's production, with six other countries producing most of the rest (see Figure 80).[8] Mining provides about two-thirds of the uranium used by the world's nuclear power stations, with the remainder coming from stockpiles, modified uranium from decommissioned nuclear weapons, reprocessed nuclear power station fuel, etc. The demand picture is increasingly complex, with significant nuclear power-building programs underway in China, India, Korea, Japan and Russia, and phase-out programs underway in several European countries. The IAEA and OECD forecast that the demand for uranium mining will grow by somewhere between 38 and 80% by 2030.[9]

---

[8]   IAEA Secretariat (2009) p. 23.
[9]   IAEA press release, June 3rd 2008.

Figure 81. Image taken by GeoEye's GeoEye-1 satellite on October 4th 2009 of a suspected uranium enrichment plant being constructed underground at Qom in Iran. Comments by IHS Jane's analyst Allison Puccioni are inserted next to the highlighted features. Credit: IHS Jane's.

Enrichment of uranium

Uranium ore found in nature consists of two main isotopes: uranium 238 and uranium 235. Only uranium 235 can fission in a simple nuclear bomb, but it is only 0.7% of the ore. Because the two isotopes are nearly chemically identical, it is very difficult to separate them. Industrial processes to enrich uranium 235 to about 90% for nuclear weapons are very energy-intensive. The plants are generally fairly large, consume a lot of electricity and reject a lot of waste heat. The gaseous diffusion process and the aerodynamic processes used in the USA and South Africa consume thousands of megawatts and are targets for thermal imagery. The more modern gas centrifuge process also consumes large amounts of electricity but can be housed in smaller buildings and made to look like ordinary industrial buildings. The suspected enrichment plant being built at Qom in Iran is entirely underground (Figure 81), making it impossible to verify its purpose.

Plutonium production and extraction

Plutonium is produced in a nuclear reactor by irradiating uranium, so many of the same steps needed to enrich uranium are required, followed by a whole additional set

Figure 82. Annotated WorldView-1 satellite image of North Korea's nuclear reactor complex at Yongbyon on August 23rd 2009 (see text). Credit: IHS Jane's; satellite image by DigitalGlobe.

of difficult industrial practices. Nuclear reactors generate a great deal of waste heat and have many characteristics that will be obvious to trained imagery analysts. They are among the easiest (relatively speaking) nuclear facilities to identify. Uranium fuel rods from the reactor are processed to extract plutonium 239 – an extremely dangerous process because of very high radiation levels. The extraction plants generate a great deal of hazardous waste. In general, the extraction process is one that can be identified by satellite imagery, especially during the construction stage.

Allison Puccioni, a senior analyst at security consultancy IHS Jane's, comments on North Korea's Yongbyon reactor facility (Figure 82): "South Korean news agencies constantly report that North Korea is rebuilding its reactor at Yongbyon, North Korea. Though there are personnel constantly seen at the facility, the cooling tower (which was destroyed in 2008 as a gesture of goodwill) remains defunct. However, there is evidence of activity at the nearby radiochemical reprocessing center."

There were 438 nuclear power reactors in the world at the end of 2008 and another 44 under construction. Ten of the 44 were started in 2008, of which six were in China. There are also 130 civilian research reactors that are of special concern because they use uranium that is enriched to weapons grade – 76 of them in Russia and 15 in the USA. These research reactors were supplied by the super-powers during the Cold War as inducements to various countries. A total of 25 countries have at least one such reactor, including Syria, Iran, Uzbekistan, Nigeria, Ghana, Belarus and Jamaica. Being research-oriented, these facilities lack the security that surrounds military nuclear sites or even commercial power stations – no armed guards, no background checks, no security requirements and no fences with intrusion alarms. The risk of terrorists obtaining weapons-grade uranium from these sites is considered serious and potentially undermines President Obama's commitment in 2009 to "a new international effort to secure all vulnerable nuclear material around the world within four years". His call "to lock down these sensitive materials" was to "ensure that terrorists never acquire a nuclear weapon". It is possible to switch these research reactors to a fuel that has little or no bomb use but the process is difficult, time-consuming and costly.[10]

As part of the worldwide trend away from oil- and gas-burning power stations, 26 licenses to build nuclear power stations were under review by the US Nuclear Regulatory Commission at the start of 2009 but it remains to be seen whether the economic down-turn will impact the enormous investment these reactors require.[11]

TYPES OF NUCLEAR FISSION WEAPONS

Gun-type uranium weapons

The concept of a nuclear fission weapon requires sub-critical masses to be pushed together into a supercritical mass. Neutrons in the supercritical mass multiply at a tremendous rate and produce a nuclear blast.

A chain reaction starts (see Figure 83) with a neutron combining with a uranium 235 (^{235}U) atom to fleetingly become uranium 236 before splitting (fission) into two smaller atoms (e.g. barium and krypton) plus some extra neutrons (three in the case of barium and krypton) and some heat. The extra neutrons can then go on to combine with more uranium 235 atoms, each of which produces extra neutrons and heat in a cascade that grows rapidly until all of the uranium is consumed or blows itself apart. A kilogram of uranium contains about 2½ trillion trillion atoms, which is why an atom bomb produces an enormous amount of heat.

Getting the sub-critical mass to become supercritical is a difficult task. If two sub-critical masses are brought together slowly, then they begin to react as soon as they are close and then fizzle out with a relatively small and very dirty (radioactive) explosion. To get a full nuclear explosion, the pieces of uranium or plutonium must be brought together very rapidly. The gun design was used for the bomb dropped on Hiroshima in 1945 (Figure 84).

In a gun-type bomb, two sub-critical masses, a "bullet" and a "target", are propelled towards each other using gunpowder in a gun barrel. When they impact, a nuclear explosion takes place and all the material is vaporized in the explosion. Gun-

---

[10]   Broad (2010).
[11]   IAEA Secretariat (2009) p. 1.

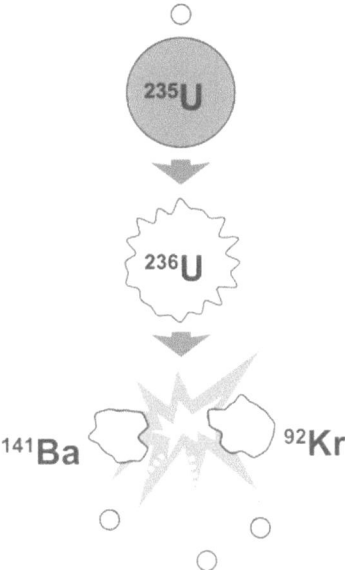

Figure 83. Uranium fission producing waste products, neutrons and heat. Credit: Wikipedia Commons.

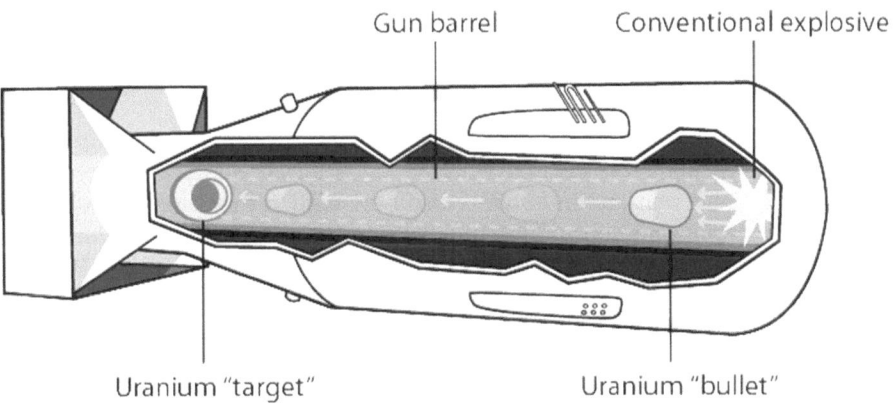

Figure 84. Schematic of a gun-type atomic bomb. Credit: Wikipedia Commons.

type bombs are inefficient in their use of enriched uranium and thus undesirable for a party who has little uranium. They are simple to construct, need almost no testing and their development is invisible to overhead imagery. Compared to the implosion type discussed below, gun-type bombs can be made relatively safe to handle. Mechanical safing features as simple as blocking the gun barrel will prevent a nuclear explosion. Gun-type devices can also be easily separated into two mechanical parts for greater safety and security, and because they use relatively low-toxicity uranium, they are a minor dispersion hazard in an accident.

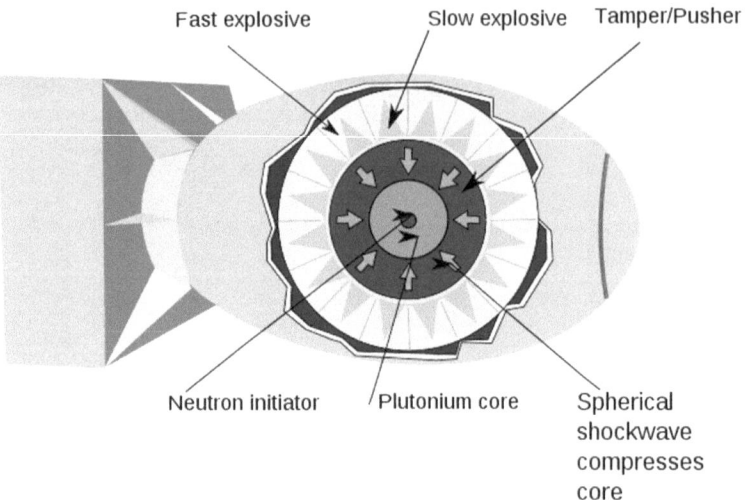

Fast explosive   Slow explosive   Tamper/Pusher

Neutron initiator   Plutonium core   Spherical shockwave compresses core

Figure 85. Schematic of an implosion-type atomic bomb. Credit: Wikipedia Commons.

Nuclear weapons are usually made from either uranium 235 or plutonium 239.[12] It is not possible to make a gun-type bomb using plutonium. Plutonium is so unstable that it begins to react before the "bullet" has completely entered the "target", thus preventing the full chain reaction from occurring.

Plutonium and uranium implosion bombs

The implosion concept was devised when it was realized that the gun-type device wouldn't work with plutonium, but it is also now used for uranium. It comprises a hollow sphere of plutonium or uranium, labeled "plutonium core" in Figure 85, that is not dense enough to be super-critical. An explosion using chemical explosives crushes the plutonium or uranium inwards (hence the name implosion) so that it becomes denser and super-critical. The conventional explosives have to be detonated with extremely high symmetry otherwise the material is expelled from the bomb – the explosives technology is called a "shaped charge". Getting the shaped charge to work correctly is extremely difficult, so that getting a large and reliable nuclear yield is hard. The explosives used in implosion bombs can be very dangerous, and if detonated, there may be a combination of some partial nuclear yield, explosive damage and the spread of highly toxic plutonium.

---

[12]   Nuclear weapons could theoretically be made from other isotopes, notably uranium 233 and neptunium 237. There are controls on these materials, but, in practice, they are not useful for weapons and are not available in sufficient quantities to be a real threat, see Bathke *et al.* (2009).

The implosion-type bomb works well for uranium or plutonium, but plutonium has a much smaller critical mass so it gives a smaller, lighter bomb and requires less high explosive.

The atom bomb dropped on Nagasaki in August 1945 was a plutonium implosion bomb. The difficulties encountered by the Manhattan Project team during 1944–1945 in developing a working implosion system have been well documented.[13] Modern computers have simplified the mathematical analysis of an implosion system, but it still remains a difficult goal – as North Korea found out when its first attempt failed to fully ignite.[14]

POLICING THE NON-PROLIFERATION TREATY: THE IAEA

Article III of the Non-Proliferation Treaty (NPT) requires every country to allow the International Atomic Energy Agency (IAEA) to verify that the Treaty is being respected. Headquartered in Austria (see Figure 86), the IAEA had been founded in 1957 as an organization to promote the "Atoms for Peace" initiative of US President Dwight D. Eisenhower. In its early years, it was focused on the promotion of nuclear power, done through outreach programs, fellowships, training and conferences. The kinds of activities promoted by the early IAEA included:

- research on radiation effects;
- publications to promote information exchange;
- establishing nuclear standards;
- assisting developing states in establishing nuclear energy programs;
- promoting international conventions on safety, nuclear waste, emergency response and liability.[15]

These tasks involved no investigations or special tools from the world of imaging. In fact, during those early days, imaging satellites were well into the future and not in any IAEA plans.

As early as 1959, a decade before the NPT existed, the IAEA began venturing into the world of international safeguards. Safeguards is IAEA jargon that covers such things as regular reporting of nuclear fuel usage and waste, analysis of samples from various locations, and scheduled and unscheduled on-site inspections by IAEA experts. For the layman, the term suggests comprehensive and foolproof monitoring, and possibly raises expectations too high. In its earliest form, safeguards were progressively applied to reactors in the developed world, such as Japan and Canada. The IAEA had no authority to monitor nuclear materials in the avowed nuclear weapons states, which initially consisted of only the USA, Russia and the UK (France and China followed as declared nuclear weapons states in 1960 and 1964,

---

[13]   Rhodes (1986) pp. 541–547, 575–578.
[14]   Collins (2007).
[15]   Fischer (1997).

Figure 86. Headquarters of the International Atomic Energy Agency in Vienna, Austria.

respectively). By 1962, the IAEA had carried out its first safeguards inspection in Norway and signed agreements to begin safeguards in Pakistan, Yugoslavia and the country now known as the Democratic Republic of the Congo.

The safeguards applied by the IAEA were rudimentary at first. They consisted of verifying that inventories of nuclear materials were as declared by member states. This was done by a process of visual inspection, book auditing and the increasing use of scientific instruments to verify that materials were in fact uranium or plutonium. In these early days, there was very little attention paid to what a state might be doing besides what it had declared. The IAEA inspectors were largely nuclear bookkeeping accountants and not detectives.

The inclusion of the verification of the NPT was a big addition to the IAEA's mission. At first, this was not particularly difficult, since most of the countries being inspected were industrial nations with professional nuclear programs. These countries had signed up in good faith and were not intending to cheat on their obligations. As time went by, more and more countries with poorly developed nuclear programs signed the NPT. Many of the new signatories were interested in joining the IAEA so they could have access to nuclear energy and several of them wanted to conceal military adventures behind the cover of peaceful programs. Joining the NPT is important for a country because without it, countries are barred from receiving nuclear assistance and are banned from international trade in nuclear materials. These agreements are enforced by the Nuclear Suppliers Group (NSG) and other organizations that are entirely separate from the IAEA.[16]

In addition, many countries signed up that had no nuclear programs whatsoever and were simply looking for the grant money that the IAEA hands out to promote the peaceful use of nuclear energy.

Soon, the IAEA was conducting safeguards in dozens of countries with relatively small nuclear programs but big ambitions. Many of them had nuclear weapons in mind, at least as a contingency. We will consider the cases of Iraq and North Korea in some depth later in this chapter but first we take a brief look at a group of countries that the IAEA has actively monitored in recent years and that are probably the most significant in the near term: India, Pakistan, Libya, Iran, Syria, Israel, Taiwan, South Africa, Brazil, Argentina, Algeria and Egypt.

COUNTRIES IN WHICH THE IAEA HAS BEEN PARTICULARLY ACTIVE

India

India is an example of a state that did not sign the NPT but used its connections in the developed world to acquire nuclear facilities. They promised not to use safeguarded facilities for weapons development but made a nonsense of the language of the Treaty by claiming that their first nuclear explosion in 1974 was a peaceful nuclear explosive and thus not in breach of the NPT. Since India had acquired a small plutonium-producing reactor from Canada and the USA for peaceful purposes only, they argued that they had not violated their agreements. This thin argument plagued India for many years until 2009, when the USA–India agreement removed US sanctions for the nuclear weapons program and allowed India to engage in international trade in nuclear materials. The IAEA and the Nuclear Suppliers Group were important parties to this 2009 agreement. The IAEA conducted inspections in India from the 1970s until 2009, forced to ignore military activities alongside civil

---

[16]   The Nuclear Suppliers Group (see *www.nuclearsuppliersgroup.org*) and the Zangger Committee (see *www.zanggercommittee.org*) govern trade in nuclear materials. The Missile Technology Control Regime (MTCR, see *www.mtcr.info*) controls trade in ballistic missile materials and the Australia Group (see *www.australiagroup.net*) is an informal group that seeks to prevent trade in materials for chemical and biological weapons.

ones by its legal mandate. India's readmission into the international nuclear trade despite developing weapons and not signing the NPT is seen by many as a serious weakening of the NPT. The IAEA will continue to verify some civilian materials in India alongside military activities for the foreseeable future.

India has an extensive satellite surveillance program – discussed in Chapter 4 (commercial) and Chapter 8 (military).

Pakistan

Pakistan, like India, has not signed the NPT to date. They have developed a massive military nuclear program that is on track to be larger than those in Britain, Israel, India and maybe eventually France. Pakistan has used the knowledge gained from safeguarded civil facilities to build their weapons program while being inspected by the IAEA. Pakistan did not explode nuclear devices until 1998, although they are believed to have had workable nuclear designs in the 1980s.

A notable thing about Pakistan is the willingness to sell nuclear technology to other proliferators. This is largely attributed to Dr A. Q. Khan, a Pakistani metallurgist who stole enrichment information from the Netherlands in the 1970s and brought it to Pakistan. He used this technology to enrich the uranium for Pakistan's first bombs, the designs for which he bought from China. Khan is given far too much credit for developing Pakistan's indigenous weapons, but his mischief in selling the technology for weapons of mass destruction to Libya, Iran, North Korea and possibly others is terrorism on an unprecedented global scale.[17]

The seemingly irreconcilable differences between India and Pakistan over control of Kashmir suggest that these two nuclear-armed countries will remain in a state of tension for years to come. Western efforts to impose peace in Afghanistan may exacerbate these tensions, since an extremist Afghanistan regime is seen by some authoritative commentators as a key part of Pakistan's arsenal in pursuit of its Kashmir ambitions.[18]

Unlike India, Pakistan lacks its own surveillance satellites, although it has launched two small prototypes – the most recent in 2001. It obtains imagery from commercial sources, which, in recent years, has included images with resolutions of better than 1 m. These sources of imagery are adequate in peace time, but it remains to be seen whether in time of tension with India, Pakistan will feel compelled to orbit a spy satellite of its own. Its fledgling space agency focuses on building up the capability to process imagery from the satellites of other countries, which, as we have seen in Chapter 4, there are now in abundance. The current close ties of the Pakistan Administration to the USA in connection with the war in Afghanistan presumably gives the Pakistani military access to at least some American spy satellite imagery,

---

[17]   Langewiesche (2005).
[18]   Rashid (2009) p. 41.

but probably not to imagery of Indian military forces. Of the significant nuclear powers, Pakistan is alone in not having guaranteed access to spy satellite information about the forces of the countries that it considers to pose the greatest threat.

Libya

Libya has had an on-again, off-again interest in nuclear weapons for decades. The country donated money and materials to others, notably Pakistan, hoping to share in any products. Libya also purchased components of a complete nuclear weapons program, but lacking a number of critical items and without a plan on how to put it all together. In late 2003, Libya was unmasked as having purchased centrifuge equipment from A. Q. Khan of Pakistan. Subsequent revelations by Libya have helped to roll up the Khan network and have brought the Libyan regime into much more favorable focus in the West.

Libya is typical of many small countries. They have a grand nuclear research center, in very poor condition, provided by the Soviet Union or the USA when gifts of nuclear technology were foreign policy tools. The center was under IAEA safeguards but none of the clandestine activities was conducted close to where the IAEA was inspecting and they went unnoticed.

Iran

Iran is the biggest problem in international safeguards today. There have been continuous developments and accusations of a clandestine nuclear weapons program since 2002. The IAEA inspects many sites in Iran, but has been unable to find any undeclared activities. Hence, the IAEA is regularly on record as saying "there is no evidence of diversion of declared nuclear materials in Iran". This should not be considered a reliable statement of the situation and can be read as "what we know about has not been subverted or diverted, but we don't know what we don't know".

It was perhaps understandable that Iran would seek nuclear weapons when American policy was to invade countries that it considered as terrorist in nature[19] and stated that Iran was one such. Afghanistan was first to receive this treatment in 2001, then Iraq in 2003. President Bush lumped Iran in with Iraq as a threat to world peace in his State of the Union speech on January 29th 2002 – probably the most authoritative forum for a Presidential statement and thus one that could not lightly be ignored. An undesirable but predictable consequence of Iran's toying with nuclear weapons is that other countries in the region are encouraged to do likewise. Countries worried about Iran's zeal in spreading its Shiia form of Islam have begun to seek assistance on nuclear energy from the International Atomic Energy Agency,

---

[19]   See, e.g., Posen (2000) p. 161.

a development that is seen by many as the first step in a dangerous nuclear arms race in the west Asia region. Nations that have taken this first step include Turkey, Saudi Arabia, Egypt, Syria and the United Arab Emirates – some of whom have the oil wealth to purchase whatever technology they decide that they need.[20]

A wild card in this dangerous situation is the possibility of Israel attacking Iran's nuclear facilities, and thereby provoking widespread violence. Many commentators find this scenario plausible because Israel has shown itself willing to take huge risks to prevent its Arab neighbors acquiring a nuclear weapons capability. The most dramatic illustration of this came in 1981 when Israeli planes bombed and destroyed the nearly complete Iraqi nuclear reactor at Osiraq near Baghdad, to prevent Iraq constructing an atomic bomb. Iraq, of course, denied that it was planning to divert plutonium from the French-built reactor to a bomb program, but even IAEA reports that no such diversion would occur failed to convince the Israelis and they sent in the F-15s and F-16s from Etzion Air Force base 1,100 km away, flying low over the semi-hostile territory of Jordan and Saudi Arabia to reach their target. In the light of Israel's willingness to initiate preventive aggression against embryonic nuclear facilities in potential adversarial countries (primarily Islamic), Iran has taken care to distribute its nuclear facilities around the country (and Pakistan has installed heavy air defense units around its nuclear facilities).

Optimists see the above considerations as a chance for Israel to make peace with the Sunni Islamic states in West Asia (that is all Islamic states except Iran and perhaps Iraq) given the common concerns of Jews and Sunnis that Iran is a threat.[21] As we will see below, Israel seems to be finding common cause with India in space, which is likely to deepen suspicions about its intentions among Islamic countries.

Iran became one of the few countries able to build its own satellites and launch them into orbit when in February 2009, its Safir rocket placed a small (27-kg) experimental satellite in orbit. A mock-up of the rocket to place a much larger satellite in orbit was placed on display a year later and it is thought this could send a satellite of up to 700 kg into space. Iran's rocket technology seems to have a North Korean and Russian heritage.[22]

Syria

Syria is an odd case in which there was no serious suspicion of nuclear activities outside of the small research center in Damascus, which is regularly visited by the IAEA. Then, in 2007, reports began circulating claiming that a nuclear reactor being built by North Korea in the Eastern Desert of Syria had been bombed by Israel. The site was first publicly identified by American researchers David Albright and Paul Brannan in October 2007. The researchers used purchasing logs from commercial

---

[20]   Broad and Sanger (2007).
[21]   Goldberg (2009).
[22]   Ben-David (2010), and Opall-Rome (2010).

vendors to determine that there was a high degree of interest in a particular imaging path in the weeks prior to the bombing. By purchasing the images central to the areas of interest, they found a bombed building that they believe is the reactor. The IAEA has been negotiating with Syria for access to this site and other locations they believe are relevant to the bombing, but only one short visit has been allowed. The IAEA continues to safeguard the small nuclear reactor and has no other noted suspicions about Syria.[23]

Israel

It is well known that Israel has maintained a policy of ambiguity about its nuclear weapons for decades.[24] It is thought to have about 200 weapons made from plutonium from their reactor at Dimona, but they have not signed the NPT, so, like Pakistan and India, they continue to operate outside the international norm. The IAEA conducts safeguards on the small Beersheba reactor in Israel but is unable to inspect the Dimona facilities, which officially do not exist.

Israel possesses both civil and military surveillance satellites, giving it assured access to imagery of countries that it considers to be of interest. A radar imaging satellite called TecSAR, launched in January 2008 via an Indian launch vehicle, gives Israel all-weather, day and night, imagery, although not as detailed as from its Ofeq optical spy satellites – TecSAR resolves objects as small as 1 m in size, whereas Ofeq gets down to 50 cm.[25] The most recent in the optical satellite series, Ofeq-9, was launched in June 2010. Israel has already sold its imaging radar technology to India for that country's Risat-2 satellite launched in 2009 and discussions are thought to be underway to build a joint Israel–India TecSAR-2 for launch in 2011.[26] The image in Figure 87, taken by one of Israel's civilian EROS surveillance satellites, is the sort of scene likely to be of interest to both Israel and India.

The Kyl-Bingaman Amendment to the US National Defense Authorization Act of 1997 requires satellite operators to degrade the resolution of data over Israel to no better than that available from commercial satellites. Israel is the only country to be granted this treatment, illustrating the influence of the Israeli lobby on US foreign and security policy. In general, the US Government can restrict access to any satellite imagery from US commercial or public sources on grounds of national security, although some critics argue that this would be a prior restraint on speech and thus an infringement of First Amendment rights. Canada, India, Israel itself and many other countries impose restrictions in the name of national security on companies that operate imaging satellites. As discussed in Chapter 1, constraining resolution and not other features of a camera (spectral and brightness discrimination, revisit time, etc.)

---

[23]   Albright and Brannon (2010), and Reed and Stillman (2009) pp. 290–291.
[24]   Norris (2007) p. 177.
[25]   Covault (2007e).
[26]   Opall-Rome (2009).

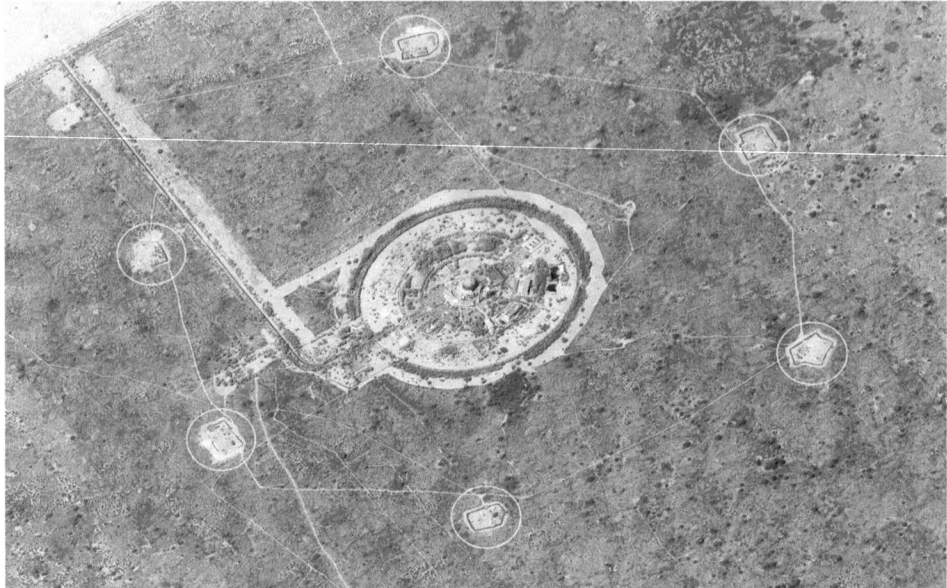

Figure 87. Israeli EROS B satellite image of Pakistan nuclear facility at Khushab. ©
2010 ImageSat International N.V., licensed by ImageSat International N.V.

leaves loopholes in the law. The Kyl-Bingaman amendment also leaves it to the satellite
companies to decide what "commercial" means – for example, is imagery from India's
IRS or Cartosat satellites commercial, given that they are owned by the Indian
Government? There are several other similarly ambiguous examples in which it is
unclear whether a satellite is "commercial" or not. Furthermore, the phrase "national
security" is vague and open to varying interpretations by different governments.[27]

Taiwan

Not once, but twice, Taiwan was caught by the USA trying to extract plutonium
from research reactor fuel. This breakaway Chinese province has been desperate to
acquire nuclear weapons for much of its existence and is reluctant to depend on the
USA for a nuclear umbrella. Taiwan has a safeguarded civil research program and
several electric power reactors. These have not been a problem for nuclear
proliferation since 1988, when the USA physically destroyed some Taiwanese
nuclear program facilities to stop them for good. Today, the IAEA safeguards the
civil activities and all is quiet. The nuclear program remnants are now under the
guidance of the Chung-Shan Institute of Science and Technology (CSIST), a highly
competent developer of arms and military systems. Will Taiwan try to develop a

---

[27]   Harris (2009) pp. 25–26, and Prober (2003) p. 2.

nuclear arsenal for a third time, this time using gas centrifuges and uranium? Only time will tell.

Taiwan purchased a small pre-operational surveillance satellite called Formosat-2 (previously called Rocsat-2). Launched in 2004, it provides image resolution of 2 m. Perhaps more significantly, Taiwan has announced plans to develop a rocket with which to launch its own satellites – initially only small and experimental satellites.[28] A country that builds its own rocket as well as its own spy satellites has real assurance that it can keep an eye on potential enemies. Russia, the USA, Europe, China, India, Japan and Israel are currently the only autonomous space powers in this sense. Brazil is likely to join them soon and other potential members of this club include South Korea, North Korea, Taiwan and Iran.

South Africa

South Africa is the only state known to have developed nuclear weapons and then voluntarily given them up. In the 1970s, South Africa was looking to add value to its extensive uranium reserves by enriching them. They developed a unique and energy-intensive aerodynamic process to do the enrichment that has only been used in South Africa. The possession of this technology, along with a repressive minority white government, proved too tempting to the hawks in the government. The enrichment program was diverted from its original peaceful purposes and turned into a military program. South Africa eventually built six crude gun-type nuclear weapons and was building a seventh. The gun type was chosen for simplicity and the fact that it didn't need to be tested before use because of the simple physics involved. South Africa carried out all of this development outside of the NPT, which it did not sign until July 1991, so they committed no breach of obligations. The weapons program went on while the IAEA was inspecting civil facilities at the same time, officially unaware of the program. South Africa destroyed their weapons just before the Apartheid era came to an end and the white government was replaced.

Brazil and Argentina

Both these countries have toyed with nuclear weapons development. Brazil has apparently gone much further, even to the point of digging a test shaft in the Amazon jungle for a nuclear device test. Brazil has also gone much further into the field of uranium enrichment than Argentina and has published papers on the complexities of nuclear device design. In 1991, Brazil and Argentina formed a joint verification program to inspect each others' nuclear facilities. This group, called the Brazilian–Argentine Agency for Accounting and Control of Nuclear Materials

---

[28] Pirard (2007).

(ABAAC), is an encouraging example of cooperation at lower than the IAEA level for transparency between two historical rivals.

Brazil has been building surveillance satellites for more than a decade. Three CBERS satellites are now in orbit, all in partnership with China. Brazil and China share the manufacturing of the satellites and China supplies the launchers. The most recent of the three, CBERS-2B, launched in 2007, is the only one with a capability that might be of interest for arms control or military surveillance – its high-resolution camera can distinguish objects as small as 2.6 m.

Argentina has built a series of satellites that can monitor maritime and biosphere parameters such as land cover and broad area pollution. Looking to the future, its SAOCOM program will include two radar imaging satellites due for launch in 2012 or 2013. SAOCOM is intended to be coordinated with the Italian COSMO/SkyMed satellites, offering a rapid-response service similar to that of Britain's Disaster Monitoring Constellation (see Chapter 5). Although COSMO/SkyMed has definite military uses, SAOCOM images will not provide as much detail as many of the current generation of commercial surveillance satellites (see Chapter 4) – SAOCOM will only distinguish objects larger than 10 m in size.

For the moment, then, neither Brazil nor Argentina has invested in true military surveillance satellites, consistent with the peaceful context of their coordinated nuclear policy.

Algeria

Imagery was credited for discovering a nuclear reactor in Algeria in 1991. The reactor was well under construction by the People's Republic of China when it was detected. Once discovered, the Algerians had no choice but to put it under IAEA safeguards. No other elements of a nuclear weapons program have been found in Algeria but this strange episode leads one to wonder what they were thinking.

Algeria has two surveillance satellites in orbit: Alsat-1 and Alsat-2A. Launched in 2002, Alsat-1 is a member of the Disaster Monitoring Constellation created by Britain's Surrey Satellite (see Chapter 5) and provides wide-area imagery with 32-m resolution, of limited relevance to arms control or military surveillance. In 2010, the first of its second-generation surveillance satellites, Alsat-2A, was launched with image resolution of 2½ m in black and white and 10 m in color. Alsat-2A has been built in France, but construction of its twin, Alsat-2B, will take place in part in Algeria under a technology transfer agreement between France and its former North African colony. The Alsat-2 family of satellites will give Algeria a significant capability when it comes to monitoring events of military significance in the region.

Egypt

Egypt never has had a visible nuclear weapons program but they always make the list of countries of concern. Egypt has the requisite Soviet-era nuclear research center at

Inshas, with a small research reactor, now in deplorable shape. They have added a larger research reactor from Argentina in recent years. The IAEA inspects a number of facilities at Inshas and there is no sign of a nuclear weapons program. In 2005, a number of reporting discrepancies were discovered that raised eyebrows. In fairness to Egypt, these were all found from indications in open literature and were likely the result of inattention to the rules as opposed to a nuclear program.

Egypt has shown no appetite to purchase surveillance satellites – its space ambitions have been focused on geostationary satellites for commercial communications and broadcasting. The most active of the potential North African and West Asian Islamic states that could have surveillance satellites is Turkey, which is buying a satellite with resolution of about 50 cm from Italy and France. The 2-ton Gokturk satellite, as it is called, is costing about $350 million and will be launched in 2011. Another Islamic country with space surveillance ambitions is Kazakhstan, which is buying two satellites from Europe, to be launched in 2014, together with an image analysis facility. One of the satellites will have 1-m resolution, which will provide imagery relevant to arms control and other security concerns. The second Kazakh satellite will have the wide-area capability of the satellites in the British-led Disaster Monitoring Constellation (see Chapter 5). The Baikonur launch complex in which Russia undertakes a large proportion of its space launches is located in Kazakhstan – representing that country's main space asset.[29]

THE IAEA AND IRAQ

The IAEA is drawn into the Gulf War of 1991

In 1991, the IAEA suffered a huge shock. The Agency had been inspecting several small nuclear facilities in Iraq and finding no problems. At the time, the inspectors could only travel to buildings designated by the state and inspect activities declared by the state. In the case of Iraq, this included a small Soviet-era research reactor, a small nuclear reactor fuel fabrication research plant and the bombed shell of another (French) research reactor. The bombed facility started out life as a heavy water[30] "research reactor" sold to Iraq by France in 1977 and was ideal for producing plutonium for weapons. Israel felt the Osirak reactor was a threat to its existence and bombed it shortly before it was to have been started. The main reactor was destroyed in the June 1981 bombing, but highly enriched reactor fuel and a research section remained and were subject to IAEA inspection.

In 1991, satellite imagery was still closely held by the states lucky enough to

---

[29]   de Selding (2009a).

[30]   "Heavy water" contains hydrogen in its rare deuterium form; deuterium is an isotope of hydrogen with a neutron and a proton in its nucleus, unlike the common hydrogen isotope, which has a proton alone.

collect it. What the intelligence agencies knew, but not the IAEA, was that the few buildings the IAEA was inspecting in Iraq were part of a vast protected site with over 100 buildings. Western intelligence knew about the site and had speculations about clandestine activities there. They also had information and images of a number of other sites in Iraq that were known to be working on an undeclared nuclear program – a program unknown to the IAEA.

The Iraq War of 1991 and its aftermath

In 1991, an alliance, led by the United States, carried out extensive bombing of Iraq. The war was sanctioned by the UN Security Council because of Iraq's invasion of Kuwait in August of 1990. A contributing concern leading to the international agreement to attack Iraq was the many indications of its serious Weapons of Mass Destruction (WMD) programs. These included medium-range missile development, chemical weapons development for use against Iran and its own people, and European purchases that were clearly for nuclear weapons. The USA, in particular, was anxious to bomb and invade Iraq to stop these WMD programs. Many of the first targets to be bombed were the nuclear facilities that the USA had monitored in imagery and via other sources. Five sites were bombed as suspect nuclear program facilities:

- Al Qaim superphosphate plant in the western desert, for uranium production;
- Al Jazeera uranium conversion plant near Mosul, for feed materials for nuclear processes;
- Tarmiya uranium isotope separation plant, partially operational;
- Ash Sharqat uranium isotope separation plant, under construction;
- Tuwaitha Nuclear Research Center, headquarters of the program and R&D (Figure 88).

Many other plants and factories that contributed to conventional and nuclear arms were bombed, but they did not handle nuclear materials. Other than Tuwaitha, which housed the two small reactors and the nuclear fuel fabrication plant, the IAEA was unaware of the existence of nuclear activities in Iraq.

The IAEA is an organization that knows its own statute of creation, and subsequent mandate from member states, extremely well. Throughout its existence, the IAEA always stayed very close to its mandate and did not investigate or question rumors of other activities. Going into the 1991 Gulf War, the IAEA only inspected the three buildings within Tuwaitha and did not investigate or question what else was going on within the vast site (Figure 89).

The IAEA had no resources to discover other nuclear activities in Iraq. The governments that did know did not share what they knew with the IAEA. Detailed newspaper accounts of Iraq's nuclear equipment-buying sprees in Europe and European aid to a uranium enrichment program by gas centrifuge went unheeded in Vienna. The IAEA was deaf and blind to most of what was going on in Iraq in 1990.

Following the bombing of Iraq, the Allies invaded the southern part of the country. The land invasion stopped quickly and an armistice was signed. The

Figure 88. Images of the destruction of the Iraqi Tuwaitha Nuclear Research Center, taken in 2003, 12 years after the bombing.

invading forces did not reach any of the nuclear sites or the chemical weapons and missile facilities that the Allies had discovered through imagery and other sources. The huge investment that Iraq had made in missiles and WMD had been severely damaged by bombing but escaped any ground attack. More importantly, no new intelligence about what Iraq had been doing was generated by the war. Many felt this was a huge mistake and so the cease-fire agreements with Iraq specified a highly intrusive post-war inspection regime in Iraq. Because the war had been fought with UN approval, the inspection teams were to be organized under the UN flag.

Resolution 687 Creates UNSCOM and the IAEA Action Team[31]

The inspection program in Iraq was initially created by UN Security Council (UNSC) Resolution Number 687. UNSC 687 created a new inspection organization

---

[31]   *http://daccess-ods.un.org/access.nsf/Get?Open&DS = S/RES/687%20(1991)&Lang = E&Area= RESOLUTION.*

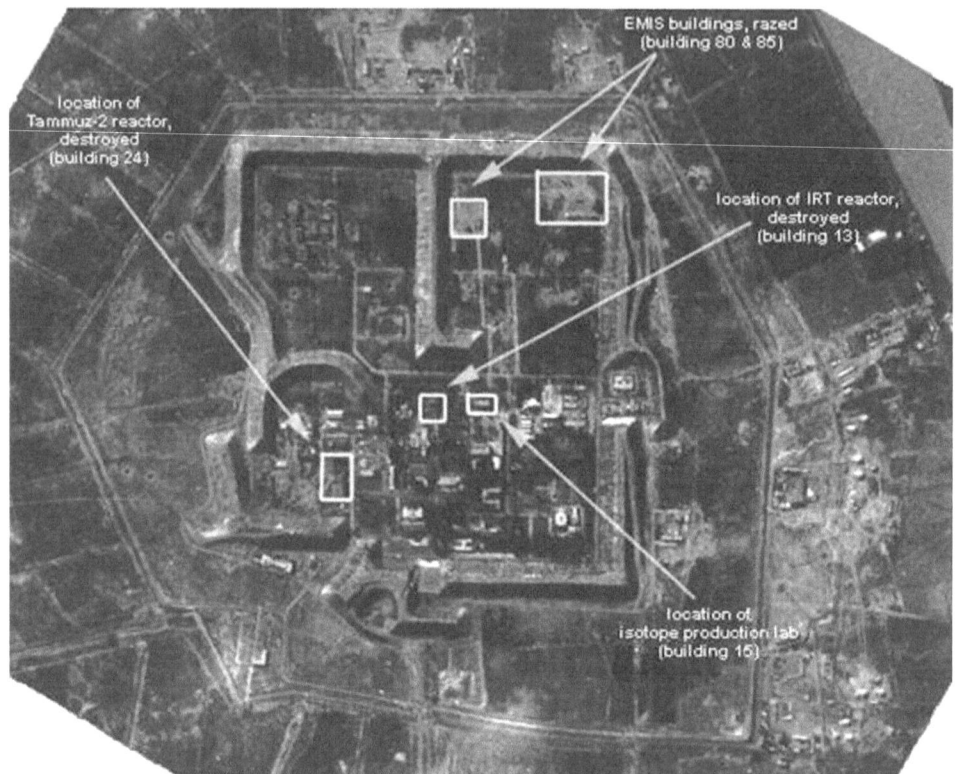

Figure 89. The IAEA was only inspecting three locations on this large Iraqi nuclear site. Note the 30-m-high earthen berms around the site to protect it from aerial attack.

called the UN Special Commission, or UNSCOM. This organization was tasked with an aggressive inspection program to discover any activities that Iraq had undertaken to produce long-range missiles, chemical, biological and nuclear weapons. Any prohibited activities discovered were to be destroyed, removed or rendered harmless. The resolution specified that the Director General of the IAEA was to carry out the nuclear portion of the mission with the assistance of UNSCOM. While the IAEA had a great deal of autonomy, it needed to cooperate with the Chairman of UNSCOM, who had some control over IAEA activities.

UNSCOM was created as a new organization that needed to acquire staff, equipment and define its mission. It recruited experts from many UN member states, with strong representation from the USA and the UK. The chairman, Rolf Ekeus, was a Swede, but his experienced deputy, Robert Galluci, was from the US State Department. There were no international organizations equivalent to the IAEA for the missiles, chemical and biological weapons task in front of UNSCOM – the Organization for the Prevention of Chemical Weapons (OPCW) was not founded until 1997. There were missile experts available, but not from a single organization. The World Health Organization (WHO) provided experts to UNSCOM, but they had no ability to stage field inspections. The IAEA was the only organization ready to quickly begin an inspection program, but only in the nuclear area.

The UN Security Council faced a problem with the IAEA at this time. The credibility of the IAEA had been severely damaged by the discovery of a nuclear weapons program in Iraq, effectively under its nose and close to facilities it was inspecting. For example, the safeguarded IRT-5000 reactor at Tuwaitha had been misused in several ways without IAEA discovery. These failures highlighted the lack of zeal that the IAEA had shown by sticking close to its mandate and not showing curiosity about undeclared activities. On the other hand, the IAEA had behaved in a legal and proper manner by sticking to its mandate and had avoided difficult political problems by its constant caution.

Outweighing this credibility problem was the fact that the IAEA had tools and experts who could immediately return to Iraq to verify the whereabouts of a significant amount of nuclear material stored there. In particular, there was an urgent need to account for highly enriched uranium reactor fuel that was unprotected in Iraq. The same team would also have to do the detective work to discover and then destroy an undeclared program, which was still only a hypothesis outside intelligence channels.

The solution to the dilemma was to create a new, stand-alone organization within the IAEA known as the Iraq Action Team. The Action Team reported directly to the then Director General of the IAEA, Dr Hans Blix, and then to the UN Security Council. The Action Team would use traditional resources to verify the uranium stocks within Iraq, but a host of outside experts were assembled to do the detective work of discovering the extent of the weapons program. The Action Team was not part of the Department of Safeguards of IAEA, the organization that had failed to discover the weapons program. The Action Team could draw on IAEA safeguards people and equipment as needed.

The Action Team was headed by a retired Deputy Director General, Maurizio Zifferero, who had not worked in safeguards. He understood that there were two separate and important missions for his team. One was to find and verify the uranium that Iraq was known to have before the war. Iraq willingly led the inspectors to the highly enriched uranium reactor fuel, which had been hidden for safety in farm water stock tanks during the bombings. Ordinary IAEA inspectors carried out this mission efficiently and professionally.

For the other mission of detecting activities not acknowledged by Iraq, member states supplied a large cadre of experienced analysts with knowledge of the country, data-gathering techniques and a wide range of nuclear skills. From the USA came technical experts from the National Laboratories, imagery analysts from Washington and political experts from the Department of State. The UK and France supplied many similar experts and a few were furnished by Russia.

Another source of expertise for the Action Team was the URENCO tripartite consortium for uranium enrichment, who supplied a handful of key experts regarding gas centrifuges for uranium enrichment. URENCO is a company owned by the governments of the UK, Germany and the Netherlands and has developed modern gas centrifuges for separating uranium isotopes beginning in the 1960s. One of URENCO's claims to fame is that in 1974, A. Q. Khan of Pakistan stole drawings and specifications for early centrifuges from URENCO and took them to Pakistan. There, he built a uranium enrichment plant that eventually supplied the uranium for Pakistan's first nuclear bombs.

By the time of the first Gulf War, the UK knew that not only had Khan built a plant in Pakistan, but he had clearly sold some of his information to Iraq. In addition, a number of Germans with past access to URENCO secrets had actively assisted Iraq with plans, specialized equipment, lists of specialized suppliers and actual centrifuge parts. Iraq was aggressively shopping in Western Europe for items unique to the stolen centrifuge plans and was approaching suppliers with highly specific requests. URENCO employees were cooperating with intelligence and law enforcement well before the war, as early as 1989, and several were on the first IAEA teams as experts. The IAEA team members were surprised to see how quickly the outside experts uncovered the clandestine Iraqi centrifuge program based upon Khan's designs. The experts knew exactly what to look for and where to look!

US Government Imagery Support to UNSCOM and the IAEA

The early days of the IAEA inspections brought about some stunning changes in the US Government's approach to security and classification of satellite imagery. The National Photographic Interpretation Center (NPIC) made up detailed maps of all the sites that the IAEA teams were going to visit. These line drawings were very accurate and showed all major buildings, fences, gates and even shipping containers (Figure 90). At first, there was some effort to keep the drawings closely held to "NATO alliance" inspectors but soon the drawings were in use by the whole teams.

A negative and bizarre side effect of the drawings was that all the buildings were numbered by the US analysts for ease of reference. Soon, the inspectors would say to the Iraqi minders "take me to Building 15" (for example). The Iraqis didn't know which "Building 15" was because it was the NPIC numbering system. When the inspectors eventually got Iraqi documents with Iraqi numbering in them, the Iraqi would not give a key to coordinate the two numbering systems and insisted on using the NPIC numbers such that some buildings were never properly identified. Eventually, the Iraqis got copies of all the line drawings used by the inspectors and began using them in their reports to the IAEA and UNSCOM.

The other exceptional release of imagery-related information by the US Government was to give near-real-time imagery briefs to the IAEA and UNSCOM. US imaging systems would image Iraqi deception activities and within a few hours, textural summaries of the information and driving directions for inspectors were with the IAEA teams in Baghdad. Of necessity, information ended up transmitted without encryption to the Holiday Inn hotel in Bahrain and on to inspectors in Baghdad. Clearly, the Iraqis might intercept these insecure transmissions, but this was offset by the ability of the inspectors to immediately respond to a developing situation.

Tuwaitha
Nuclear Research Center

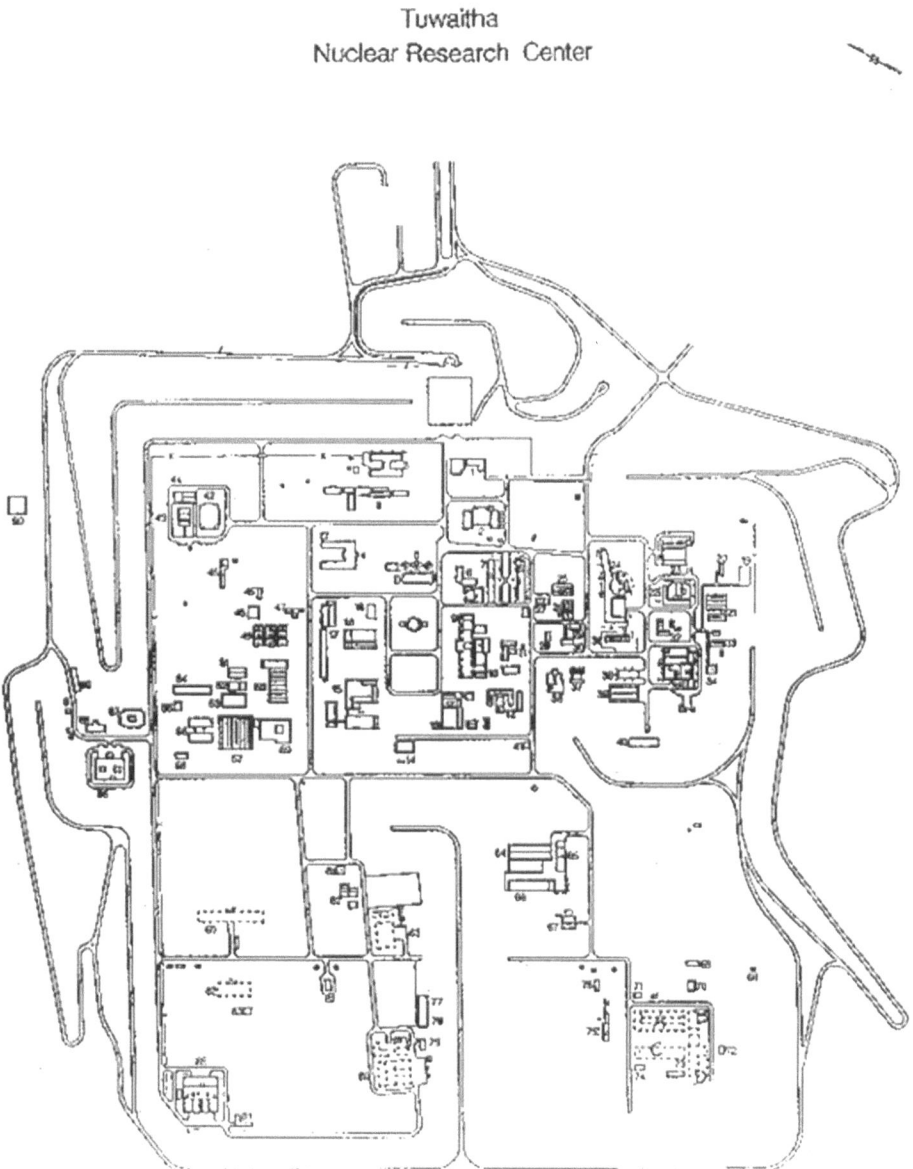

Figure 90. This is an example of the highly detailed line drawings given to UNSCOM and the Action Team by the NPIC. These drawings were considered very sensitive at the time because the high-resolution satellite imagery from which they were derived was still very classified.

Figure 91. Action Team inspectors examine an electromagnetic isotope separation machine used to separate highly enriched uranium from natural uranium. Photo Credits: Action Team 1991-1998/IAEA.

The most famous such incident is when Iraq was carrying off massive pieces of uranium-contaminated debris in July of 1991 from a large enrichment plant that had been bombed that January. Surveillance noted that huge amounts of equipment had been removed before an IAEA Action Team inspection, placed on a long convoy of trucks and taken to a military base. The IAEA responded to the base where the trucks could be seen departing from a rear exit. The Action Team stopped the convoy on the highway and established the first solid proof that Iraq had a secret uranium enrichment program, never declared to the IAEA, as was required (Figures 91 and 92). A young American Army Captain, Rich Lally, took the incriminating ground photos of the Calutron enrichment machines that confirmed the satellite imagery. The Iraqis surrounded him and demanded he turn over his camera and film. Lally refused, remarking later that his wife would have killed him if he had lost *her* camera!

This bold use of classified imagery, instant declassification and briefing to non-US nationals to get the job done was unprecedented and highly effective. The net result was that perishable intelligence was used as needed and neither sources nor methods were seriously compromised.

Figure 92. Inspectors examining uranium separation equipment that was being removed by Iraqis and spotted by satellite imagery.

Destruction of facilities in Iraq

One of the tasks facing the Action Team in Iraq was to destroy or render harmless any facility or equipment that had escaped the bombings of January 1991.

For example, the large nuclear weapons development laboratory at Al Atheer had not been detected by the USA prior to the Gulf War and had largely escaped damage. Only a few buildings had been hit because the site was unknown to the USA prior to the post-war inspections and it did not have unique features ("signatures", as the analysts call it) that would cause it to be targeted. It looked like an ordinary factory. The worst damage was to an instrumentation bunker, because bunkers protecting equipment and personnel were an automatic military targeting priority. This bunker was specially designed to protect experimenters and their sensitive recording equipment from conventional blasts of explosives on a pad outside. Precise measurements of the speed and symmetry of a spherical implosion are necessary to develop a nuclear bomb. It was ironic that this building was probably targeted because it was thought to be an explosives store when, in fact, it was a key scientific research building in the nuclear weapons program.

In April of 1992, the Action Team directed Iraq to destroy eight buildings at Al Atheer. They also destroyed electrical power facilities at the enrichment plants as well as hundreds of pieces of specialized manufacturing equipment and centrifuge parts.

Figure 93. "Before and after" – destruction of a building at Al Atheer, summer 1992, under the supervision of the inspection team. The building was used for Iraq's secret nuclear weapons program. Credit: Action Team 1991-1998/IAEA.

There were several reasons for this activity. The first was that Iraq refused to give detailed descriptions of the buildings. The Action Team by then had seized thousands of pages of documentation about Iraq's clandestine programs, in particular the role of Al Atheer. Iraq had admitted to a clandestine undeclared uranium enrichment program and other activities that could be claimed to be for civilian purposes. But, in an attempt to save face, Iraq still had not confessed to any nuclear weapons activities by 1992. Given Iraq's refusal to admit these obvious transgressions, destruction of facilities that were clearly for uranium metallurgy, high explosives development and testing in direct support of a nuclear explosives program was therefore appropriate.

In addition, UN resolutions authorizing the inspections were apparently subject to a time limit and neither UNSCOM nor the Action Team knew when they might be brought to a halt. As it turned out, inspections continued for 10 more years but at the time, the need to neutralize the nuclear weapons capabilities seemed urgent.

The first overhead images of Al Atheer came from an unlikely source. Inspectors making the regular journey to and from the UN staging base in Bahrain noted that the flight path always passed directly above Al Atheer and several other key Iraqi WMD facilities. The team negotiated with the pilot of the German army C-160 Transall cargo aircraft to do some aerial photography. Inbound to Habbaniyah airport outside Baghdad, the pilot reduced altitude and allowed an IAEA photographer to be outside the fuselage on the tailgate of the cargo plane. Blowing in the wind in a safety harness, he took the first UN images for interpretation. The images turned out to be blurry and quite useless, but the first small step in a UN-organized aerial measurements program had been taken.

The destruction of facilities began in April 1992. Iraq army demolition teams began mining the columns of the largest industrial buildings at Al Atheer with destruct charges. When the charges were fired, the buildings collapsed in an orderly pancake fashion, consistent with the work of a professional demolition team (Figure 93).

A week after the team returned to Vienna, there was a frenzied call from the USA. "We thought you had blown up the buildings at Al Atheer?" they asked, "But the buildings are still there!" Naturally, the IAEA team argued that the buildings were totally destroyed. They had witnessed it themselves.

Several American representatives then hurried to the IAEA in Vienna to view ground-level photographs of the buildings before and after demolition. They were stunned to see that the several-story industrial buildings had pancaked so neatly, that the roofs still looked the same to the imagery satellites!

Using imagery in the field

On several occasions, the USA bombed Iraq after the Gulf War in response to provocations. In one such event, in January 1993, the USA attacked a plant in the Baghdad suburb of Zafaraniyah using 44 cruise missiles. The plant was heavily damaged. It was not a plant handling nuclear materials, but, instead, it had been manufacturing mechanical equipment for uranium enrichment factories before the war of 1991. The plant was no longer engaged in nuclear activities in 1993 but it was a symbol when the USA needed a facility to attack. Action Team inspection teams had visited the plant a number of times and had noted the most critical machining and inspection equipment in the facilities. Most of the critical equipment was surgically destroyed in 1993 by the precision cruise missile targeting. This led to strong complaints from the Iraqis about the use of Action Team GPS equipment to pinpoint potential targets. The manager of the Zafaraniyah plant was under-standably furious about the damage to his plant. His only retort to the next inspection team was "You used 44 cruise missiles at about $1 million each and did about $5 million worth of damage. Your methods are not cost effective!"

In addition, the inspectors were constantly receiving tips from outside sources for their inspection program. An example was when a national organization insisted that there was a suspicious building in the security guard quarters at a former nuclear site. The Action Team did not have its own imagery at the time, so they were given a briefing and a map and sent to investigate. The suspicious building proved to be a volleyball court for the guards. In this case, and many others, the lesson learned is that headquarters analysts need to get into the field as part of the inspection process.

There is also the need for the right expertise and often it is not on the prime topic, such as nuclear or chemical weapons. The Badush Dam is a huge, partially finished structure near the northern Iraqi town of Mosul. The dam drew a lot of attention in 1993 because of rumors that nuclear work was being done in the area. The massive unfinished piers of the spillways and powerhouse seemed suspicious to nuclear engineers and many features were noted in imagery. The concern was that large industrial activities, such as a nuclear reactor, could be concealed inside a huge construction project, such as a dam, that never seemed to be finished. On one occasion, a representative from a friendly government told the Action Team that their reconnaissance pilots were bored because they didn't have enough tasking. The Action Team suggested Badush Dam as an interesting target. The pilots obviously loved the task because, soon, a bag of aerial reconnaissance photos arrived. Many of the photos were looking up at the construction cranes, indicating some very low flying!

When the images were shown to an experienced engineer from the US Bureau of Reclamation, he explained all of the "suspicious" features to the Action Team. He believed that everything was quite normal for such a huge hydroelectric dam and he led a ground inspection that confirmed it.

Credit must also be given to the analysts in headquarters, who do their job well. Amateurs tend to request the highest resolution and tend to look closely at their target. This included US inspectors, who missed the two enormous high-voltage power lines crossing farmers' fields to the Tarmiya uranium enrichment plant near Baghdad. The lines ended at a huge substation about 1 km from the factory and went unnoticed by some inspectors. It can be noted, however, that the huge electrical supply and the high security of the modern, recently built site led targeting planners to attack the site in 1991 before even establishing its purpose.

Olive Branch imagery

Early in the inspection process, the USA and UNSCOM agreed that aerial imagery was an extremely useful tool for the Iraq mission. They signed a memorandum of agreement that the USA would supply a US U-2 high-altitude imagery aircraft to the UN. It would be operated by the USA with targeting requirements provided by UNSCOM and the Action Team. The USA had been carefully showing other overhead satellite imagery to the organizations, but only for briefing and viewing purposes, and not to retain or interpret. The low-classification aerial imagery was an excellent solution to allowing UN organizations to store and examine images carefully.

The new U-2 program was christened Olive Branch and continued for many years, at least until 1999. The IAEA could give tasking requirements to UNSCOM and receive images within days. This allowed the Action Team to begin to analyze known sites in detail and to task imagery for other sites of concern. The images were only hard-copy positive prints, so the analysis was relatively primitive but extremely welcome for its day, before the general availability of commercial satellite imagery.[32]

---

[32]   Avenhaus *et al.* (2006) pp. 268–269.

UNSCOM Aerial Image Team

In addition to Olive Branch, UNSCOM started its own Aerial Imaging Team (AIT) in 1993. The AIT used helicopters provided to UNSCOM for transport within Iraq. On transport missions, or on special missions, the helicopters carried professional photographers with high-quality hand-held cameras. The helicopters were able to fly at low altitudes and hover if necessary. Images could be captured at very low angles such that building details could be analyzed almost as from a ground photo. Initially, the helicopters were provided by a German army support team. Later, the Chilean Armed Forces provided aerial support. The UNSCOM and IAEA teams in Iraq had a great deal of freedom. In theory, they could go anywhere at any time, although these rights were steadily eroded by the Iraqis. One area in which UNSCOM compromised with Iraq was on aerial photos from their own helicopters and planes, not counting the U-2. The U-2 irritated the Iraqis because they strongly suspected that the USA was taking more images than just the ones requested by UNSCOM. Therefore, they used all their bargaining power to get agreement that there would be no casual photos from the helicopters by regular inspectors.

Such agreements, fortunately, could be modified by logic. On one occasion, a team made an aerial inspection with no cameras allowed in a remote region. When the chief inspector requested a few photos to show that there was nothing at the suspect site, the Iraqi said "No". "Fine," answered the inspector. "Prepare a land mission by jeep and other transport into the remote region so we can take some photos showing there is nothing there." Reason prevailed. The photos were taken and everyone saved a lot of time.

The AIT eventually acquired thousands of images for UNSCOM and for the Action Team. The sheer volume of images highlighted the next real concern: data management.

UNSCOM becomes UNMOVIC

In December of 1999, the former UNSCOM was dissolved and replaced with UNMOVIC. The reasons were complex and political. The USA and the UK had held key positions in UNSCOM for its entire existence. The broader UN community saw UNSCOM as an American-dominated tool. There were also allegations that intelligence agencies had infiltrated UNSCOM and were using UNSCOM monitoring tools for their own purposes, including planting hidden bugs in authorized equipment. The new organization reduced the number of Americans in key positions and removed dependence on American assets. One positive outcome of this was increased reliance on commercial satellite imagery instead of classified materials shared by governments. Governments share when it suits them. UNMOVIC's new satellite imagery analysis unit could choose its own targets and its own dates of imagery. UNMOVIC and the Action Team set up broad commercial contracts with imagery suppliers and began to analyze their own imagery.

In 2002, the USA began to organize an invasion of Iraq, in response to the attacks of September 11th 2001. UNMOVIC and the IAEA had barely a caretaker role from 1998 until this change due to the fact that Iraq had expelled inspectors. Iraq felt that the inspections had discovered all that there was to find and that continuation was a political state of affairs, not a technical one. UNMOVIC was by this time headed by Dr Hans Blix, former head of the IAEA during the earlier Iraq inspections after 1991. The IAEA was now headed by Dr Mohammed ElBaradei, and his leader of the Action Team was Dr Jacques Baute, a Frenchman. The role of Dr Blix, first as IAEA head and later as UNMOVIC head, causes confusion about who was in charge of the nuclear portfolio in 2002 and 2003. Officially, it was the IAEA and ElBaradei, who owned this responsibility. The IAEA was almost independent of UNMOVIC but the latter's Chairman, Dr Blix, still had some administrative and logistical authority over the IAEA.

IMAGERY IN SUPPORT OF VERIFICATION IN NORTH KOREA[33]

Iraq was by far the biggest use of overhead imagery for the IAEA, but North Korea was also a test case for imagery analysis. In 1993, the USA approached the IAEA with many concerns about North Korea triggered by strong suspicions that this isolationist communist state had used a small electricity-generating research reactor to produce plutonium and was separating it.

An important piece of the puzzle was a site at which wastes appeared to have been buried and concealed and at which a surface-to-air missile battery had been built. The highly classified imagery was shown to the IAEA Board of Governors and then taken away. Because there was no aerial imagery of North Korea, and satellite imagery was still highly classified, the IAEA could not retain the materials.

The satellite imagery was the basis for a request by the IAEA to visit the suspect site. Access to the waste would have provided strong direct physical evidence of reprocessing activities to produce plutonium for weapons. Nuclear waste products can even be carefully dated by the decay of the many different isotopes in the waste and the signatures would be absolute proof of undeclared activity. Naturally, North Korea had not declared such activities and was very unwilling to provide access. By claiming the waste site was a military facility, they stonewalled the IAEA indefinitely.

The story of verification in North Korea has been a chequered one. The IAEA managed some inspection activities over the years but was frequently barred from inspecting obvious violations. In 2002, North Korea expelled IAEA inspectors and then in 2003, announced that it would withdraw from the NPT – an action that had many legal complexities.[34] In fact, North Korea came within 1 day of

---

[33]   The Democratic Peoples' Republic of Korea is more commonly referred to as North Korea and this latter name is used in this book.

[34]   Is withdrawing from a treaty unilaterally a "breach"? The IAEA says North Korea cannot withdraw but they say they have. North Korea no longer agree to any safeguards, which would be a breach, if they truly cannot withdraw!

Figure 94. North Korea's Unha-2 rocket on the launch pad at Musudan-ri imaged by GeoEye's GeoEye-1 satellite on April 5th 2009, just 20 min before the failed launch of a test satellite. The vehicles parked in the shadow of the rocket are said by IHS Jane's image analyst Allison Puccioni to belong to Kim Jong Il and his cadre, who then moved to the Command Centre. Credit: IHS Jane's.

leaving the NPT but put their decision on hold, only to withdraw in 2009 for good.

IAEA inspectors did not return until 2005. In October 2006, North Korea conducted a nuclear test. The test was of very low yield, so low that most observers call it a failure, but nuclear debris was detected on the winds and North Korea has to be added to the list of de-facto nuclear weapons states. In May 2009, North Korea announced another slightly larger nuclear test that was detected by seismic stations. Strangely enough, no radioactive debris has been detected from this test. Most experts believe the release and detection of radioactive debris should accompany such a test so its true nature is in doubt.

Again, in 2009, IAEA inspectors were asked to leave North Korea and that country apparently resumed plutonium extraction from old reactor fuel. The situation in North Korea continues to see-saw in one of the most politically charged and repressive regimes in the world. Surveillance by U-2 or SR-71 high-flying aircraft is probably out of the question because of the potential for a major political incident. North Korea reacts strongly to any provocation it perceives and violation of airspace would be a major one. Orbiting overhead imagery will continue to be one of the few useful windows into this proliferant state (for an example of this, see Figure 82).

North Korea claimed to have launched an experimental satellite into orbit in April 2009, but the satellite ended up in the Pacific Ocean (see Figure 94). The launch was probably timed to prevent South Korea gaining a propaganda coup with its first satellite launch that took place 4 months later, and was also a failure. The April event showed that North Korea possesses most of the necessary rocket technology and is predicted to succeed before long. A surveillance satellite is likely to be a high priority once the launcher technology is working.

South Korea is also investing in surveillance satellites – initially commercial ones. Kompsat-2 has been providing imagery of 1-m resolution since its launch in 2006. The Kompsat-5 imaging radar satellite is due to be launched in 2010 and benefits from the radar technology of Italy's COSMO/SkyMed satellites. South Korea has been developing its own launcher with assistance from Russia. Its first two attempts to place a satellite in orbit failed in August 2009 and June 2010 due to problems with the rocket.

IMAGERY AND NUCLEAR VERIFICATION

Managing large imagery files

It is worth noting that all of the imagery that was being acquired by UNSCOM in 1992 and the following years was received and stored as positive prints, although the AIT could produce duplicate positive transparencies. There was one light table[35] in New York and none in Vienna. So the interpretation of imagery was rather crude by contemporary standards. The digital world did not intrude on the Action Team until closer to the year 2000.

All of the Olive Branch images were considered classified and were stored under security terms approved by the USA. One requirement was that no image could be photocopied or stored on electronic media. Only the original hard copies were allowed. As the file grew to hundreds of images, the cataloging and security of the files became a large administrative burden. This may not seem obvious to people familiar with large government operations, but for the UN, the burden of trying to maintain control and also have easy access to the data was eventually almost overwhelming. This does not take anything away, however, from the fact that high-quality imagery was one of the best independent tools that the UNSCOM and the IAEA had for many years.

Thermal and multi-spectral imaging

Thermal imaging is a very powerful way to examine large targets that produce an excess of heat, such as reactors or gaseous diffusion uranium enrichment plants. If

---

[35] A glass-topped table lit from below.

Figure 95. Aerial photo of the massive damage to Unit 4 at the Chernobyl Nuclear Power Station following the 1986 accident.

the plant discharges waste heat to a closed body, such as a lake, there can even be fairly accurate quantitative estimates of the amount of heat discharged. These waste heat estimates can be translated into mathematical models of uranium or plutonium production.

One such estimate was done in reverse for the accident at Chernobyl in April 1986 (Figure 95). Even low-resolution Landsat images showed the sudden disappearance of the thermal plume in the closed heat sink reservoir. Of course, panchromatic imaging showed massive destruction of the reactor itself, but the thermal plume resumed in time showing that the other three undamaged reactors at the site were once again operational!

Multi-spectral imaging and even hyper-spectral imaging (see next chapter) were of use to the Action Team assessing Iraq. The team was fortunate enough to have some limited expertise on staff and benefited greatly from a grant from the Canadian Nuclear Safety Commission (CNSC). The CNSC funds a number of extra-budgetary activities at the IAEA and this is an area in which they had interest and experts. CNSC is a very forward-looking organization with a sincere commitment to the use of technology to solve nuclear proliferation problems. Their studies of the Ranger uranium mine in Australia are an example of on-going interest in this area.[36]

An early case of thermal analysis involved an underground plant in Iraq. All

---

[36] Leslie *et al.* (2002).

underground facilities in Iraq were of high interest because of their hardness against attack and the difficulty to assess what was going on inside. Inspectors could not enter Iraq in 2002, but they were interested in the status of an underground power station north of the city of Mosul in the Kurdish region. They knew that the station had been built as a pumped storage station by a Swiss company for Iraq. Water was pumped from the reservoir behind nearby Saddam Dam and power plant to the top of a nearby mountain. This happened during the night when power demand for the dam's hydroelectric capacity was low. The water ran back down through the generators to make extra electricity during the day when demand increased.

The Action Team acquired Landsat imagery of the site in early 2002, prior to restarting inspections. Sure enough, the water on top of the hill was cold, indicating that it came from the bottom of the reservoir. There was also a cold plume of water visible during the day at the foot of the mountain where cold water was re-entering the Tigris River. This finding convinced inspectors that the crowded underground facility was probably operating as designed, as a power plant, and that rumored uranium processing in the underground plant was unlikely. This was later confirmed on the ground when inspectors returned in late 2002.

An even more aggressive use of imagery used hyper-spectral sensors to study possible uranium production plants. Iraq produced a great deal of uranium by extracting it from phosphate fertilizer at the Al Qaim superphosphate plant near the Syrian border in the western desert. The uranium extraction unit of the plant was procured from Belgium and was known to American bombing targeteers in 1991. It was one of the first targets to be bombed in Iraq. The inspectors knew in 2001 that the unit at Al Qaim was destroyed, but they also knew that Iraq was continuing to mine phosphate fertilizer containing uranium. Was the uranium being extracted elsewhere?

The CNSC set up a project to characterize the hyper-spectral signatures of the phosphate fertilizers from Al Qaim and the mine at Akashat.[37] The team used the Hyperion Hyperspectral sensor in the US EO-1 satellite. The work was of a preliminary nature, but it showed that it was possible to identify the characteristics of a plume of white dust around a phosphate fertilizer plant. If a suspect site at another location was found, its dust could be remotely monitored from space to see whether its characteristics were the same. This could lead to other remote sensing studies using imaging or spectral data, and be a trigger for ground-based inspections if they were allowed.

The studies of spectral signatures from known uranium ore processing plants could be used to quantify the signatures from other plants, like Al Qaim, and possibly eliminate other plants with white powder around them, such as cement plants. This is a forward-looking way to try to use the best modern technology to solve remote sensing problems.

---

[37]   Satellite Hyperspectral Imaging In Support Of Nuclear Safeguards Monitoring.

The future of imagery in verification

UNMOVIC was dissolved after the second Gulf War of 2003. Inspections under UNSC 687 and following resolutions ended. The American Iraq Survey Group confirmed the findings of UNMOVIC and the Action Team that there were no programs to produce WMD in Iraq between 1991 and 2003. UNMOVIC's role was complete but the IAEA continues to have a worldwide mission to monitor nuclear programs around the world and to search for undeclared activities.

After Iraq, the knowledge and staff of the Action Team (called INVO in its final days) were absorbed into the Department of Safeguards. The experiences of Iraq, North Korea, South Africa and others led all advisors to the IAEA to urge it to be more curious, show more initiative and to expand its powers. This process is known as Strengthened Safeguards and is heavily dependent on information, especially from overhead resources (aircraft and satellites).

The IAEA Department of Safeguards has established a Satellite Imagery Analysis Unit (SIAU) that monitors open-source information.[38] The unit is growing, in staff and equipment. It represents one of the best hopes that the IAEA can expand its role in NPT verification.

SIAU produces high-quality site maps for inspectors to take to the field. The maps are compared with information on the ground. If there are questions, the Agency asks the site operator to explain.

The unit can also be tasked to look at undeclared activities when there is a suspicion that something is wrong. This process is delicate because the mandate of the IAEA is weak in directing the Agency to probe beyond its limited powers. A bloc of member states has asked the IAEA to stop using satellite imagery altogether, or at least to inform the state in advance that imaging will take place. This rather naive view of imagery indicates the poor understanding that many states have of imagery acquisition and sales, but it represents a serious threat to an important and efficient new tool. The IAEA is a UN-affiliated organization, and is very sensitive to any suggestion that the world body is "spying".

The SIAU also illustrates the costs of such an operation. Commercial imagery is not free. Archived scenes are relatively inexpensive, but asking for cloud-free imagery acquisition on a priority basis can be very expensive. An organization like the IAEA can never hope to have the frequency of collection or the number of sites covered that a national collection program can afford. Management of thousands of large digital files is expensive as well, leading to hardware costs and administrative personnel.

Expansion beyond panchromatic imagery to radar imagery and hyper-spectral are goals, as opposed to near-term tasks. The training and modeling required to interpret this type of data consume huge resources within governments, and beyond those of a small player like the IAEA. Hopefully, these tools will be developed aggressively for land-use studies, for example, and the results can be passed on to the

---

[38]   IAEA Secretariat (2007).

Figure 96. Secretary of Defense Robert Gates flanked by Chairman of the Joint Chiefs of Staff Admiral Mike Mullen (left), Secretary of State Hilary Clinton and Secretary of Energy Stephen Chu conduct a press conference at the Pentagon on April 6th 2010 on the *Nuclear Posture Review*. The review commits to "de-MIRVing" all US long-range missiles and gives a guarantee not to use nuclear weapons against non-nuclear NPT states. DoD photo by Cherie Cullen.

IAEA. The April 2010 *Nuclear Posture Review* report by the US Government recognizes "that the IAEA currently lacks sufficient resources and authority to carry out its mission effectively" and promises "rigorous measures to reinvigorate the NPT" (Figure 96).[39] The SIAU may therefore receive additional resources before too long.

One cannot touch this topic without mentioning the role of Google Earth and the equivalent services from Microsoft, Yahoo, etc. High-resolution data files from commercial vendors are best processed with excellent software managed by experts. These enormous commercial files come with copyright restrictions and must be properly maintained. As a result, only the cadre of specialized analysts in the SIAU normally does detailed analysis. Google Earth is a tool for the masses, including inspectors, who will only use it occasionally. It is a tool that stimulates curiosity and situational awareness. It can introduce inspectors to a whole new view of the countries they monitor and encourage them to learn more about the environment

---

[39]   Gates (2010) pp. 4, 46.

they are studying, not just accounting numbers when verifying nuclear fuel. Additionally, there must be training to ensure that there is a journeyman's approach to image analysis. Blogs on the internet today, examining sites such as the Syrian reactor bombing, show how amateur imagery analysis of the conspiracy theory school can reach the point of ridiculousness and undermine conclusions, not support them.

The IAEA is investing to improve its effectiveness and efficiency by developing and implementing new technologies and methodologies, such as special on-site inspections, unmanned in-situ monitoring, environmental sampling and open source information. Satellite imagery is increasingly part of this mix, summarized as follows in a recent IAEA status report: "In 2005, the IAEA acquired and analysed satellite imagery on a regular basis in support of its safeguards activities. Hyper-spectral imagery, which was used for the first time in 2005, demonstrated the potential for significantly improving the IAEA's ability to monitor uranium mining and milling activities. In cooperation with some Member States, radar imagery processing has been developed and partially implemented; this technique further improves the IAEA's ability to identify specific activities including activities underground."[40]

A recent review of satellite imagery and nuclear verification summed up both the strengths and limitations of the technology: "Images acquired from civil or commercial earth observation satellites are an important open source for the increasingly information-driven IAEA safeguards. The main applications of satellite imagery are to verify the correctness and completeness of the member states' declarations, and to provide preparatory information for inspections, complimentary access and other technical visits. If the area of interest is not accessible, remote sensing sensors provide one of the few opportunities for gathering data for nuclear monitoring. However, satellite imagery is far from being sufficient to solely confirm the existence or absence of nuclear activities."[41]

---

[40] IAEA Secretariat (2006) p. 4.
[41] Niemeyer (2009) p. 34.

8

Military imaging satellites: long-range intelligence

INTRODUCTION

Surf the web to Google Earth or Microsoft Bing and you can zoom in on anywhere in the world from above. A few places are blurred or pixellated like the X-rated bits on TV – your web search engine will take you to websites that specialize in identifying the censored areas if you search on something like "censored areas in Google Earth". One famous blurred area during the Administration of President George W. Bush was the home of Vice President Cheney at the US Naval Observatory in Washington, DC. As described in Chapter 1, once the Obama Administration took office, the blurring was removed.

Of course, the imagery is old. Its age varies but typically is a year or two old, so the latest deployment of troops or Air Force planes won't be evident. Military planners would like to see where their adversary's forces are at all times, so since the dawn of the space age, satellites have been used to address that. Google Earth is of interest to military planners simply because it provides mapping information of pretty much the whole world at your fingertips. Iraqi insurgents gained some advantage from that convenience, as we saw in Chapter 1, but getting serious military information takes more than that.

The military and their intelligence colleagues want current information – "What is over that hill or round that corner?" But they also want the big picture – "How many tanks or bombers or missiles or soldiers has my enemy?" Satellites can help answer both of those questions.

In this chapter, we look at what imaging satellites can and can't do to help the military and the intelligence services. In the next chapter, we will look at a related type of surveillance from space, that of electronic eavesdropping. Another military space activity, satellite navigation, was discussed in Chapter 6 and yet another, weather satellites, in Chapter 2.

Table 4 summarizes the status of the main types of military imaging satellites, classifying them as optical, radar imaging and missile alert. Each of these types is

P. Norris, *Watching Earth from Space,* Springer Praxis Books,
DOI 10.1007/978-1-4419-6938-5_8, © Springer Science+Business Media, LLC 2010

Table 4. Status of military imaging satellites and associated data relay satellites.

	Optical	Radar imaging	Missile alert	Data relay
China	Operational	Operational	Defunct	Prototype in orbit
France	Operational	Defunct	Prototype in orbit	Defunct
Germany	Defunct	Operational	Defunct	Defunct
India	Civil/military dual-use	Under development	Defunct	Defunct
Israel	Operational	Defunct	Defunct	Defunct
Italy	Defunct	Civil/military dual-use	Defunct	Defunct
Japan	Operational	Operational	Defunct	Civil/military dual-use
Russia	Operational	Defunct	Operational	Operational
UK	Prototype in orbit	Defunct	Defunct	Defunct
USA	Operational	Operational	Operational	Operational

Legend

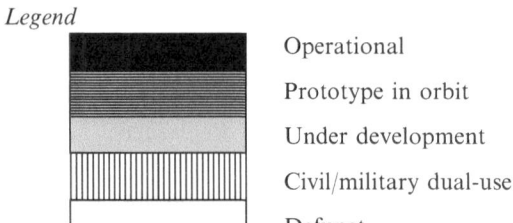

Operational

Prototype in orbit

Under development

Civil/military dual-use

Defunct

discussed in this chapter plus the special satellites that relay the images back to the home country – data relay satellites – the rightmost column in Table 4.

TACTICAL SURVEILLANCE

With the ending of the Cold War, America found itself drawn into regional conflicts in several parts of the world. The first major example was the Iraq invasion of Kuwait in 1990, which led to the Gulf War of 1991 – Operation Desert Shield, as the USA called it. Satellite imagery was in constant demand by the forces in the Gulf region to assist in their immediate actions – to tell them where enemy forces were located or to clarify the status of infrastructure such as bridges.

But although America had several spy satellites in orbit and collected a great variety of images, the forces on the ground found it difficult to get the information they wanted. Frequently, they received loads of images but didn't have the photo-interpreters to work out answers to the questions they faced.

The organizations and infrastructure set up to get satellite images to Washington, DC, where they could be analyzed by specialists supported by banks of super-computers, worked well for the Cold War strategic era. Getting information about local conditions to forces scattered across a far away desert was another matter. America spent the 10 years after the first Gulf War addressing this problem, so that by the time of the Iraq invasion in 2003, US forces were better served by the imaging satellite community. In fact, in between the two Iraq wars, the USA introduced no new weapons or surveillance systems – all of the enormous procurement defense budget was spent on completing systems like GPS that were only partially deployed for the First Gulf War and making things work better together – Army, Air Force, Marines and Navy sharing information, for example. The result was that the information reaching the troops in Iraq in 2003 was much improved.[1]

Ironically, the very success of spy satellites has led to expectations often being unfulfilled. The following anecdote illustrates the sort of mismatch between hype and fact that can arise.

A squad commander far from base in Iraq or perhaps Afghanistan calls up his intelligence officer (probably via satellite, incidentally) and asks for an immediate satellite image of the valley ahead of him. The intelligence officer has the wit to query this expensive and probably impossible request (it is highly unlikely that a suitable satellite will be in just the right place at the right time) and asks why the image is needed. The commander explains that he wants to cross the gorge ahead but is worried that the bridge across it has been bombed. The intelligence officer reminds him that even if the satellite image showed the bridge to be down, he might be able to cross if the sides of the gully were not too steep. The commander suggested that this information could be obtained by analyzing a satellite image and measuring the slope of the gully sides. The intelligence officer had a better idea – he contacted a special forces unit that he knew was in a position to look down on the valley. The special forces man looked through his binoculars at the gully and reported that cars and trucks were being driven across it by local people, having first bull-dozed a path down the sides. This information was, of course, exactly what the squad commander wanted to know.

Whereas satellites are excellent at giving the big-picture summary of Russian or American strategic forces, they are less good at giving black-and-white answers in one particular scene at one particular time. Part of the process of gauging the strategic picture is to examine the changes in a facility or location over time – years, rather than days. The tactical requirements of forces engaged in regional wars are much more immediate and explicit: Is there a tank ahead of me? Is there an enemy soldier in that ditch?

---

[1]   Kehler (2003).

KEYHOLE: US OPTICAL RECONNAISSANCE SATELLITES

Since the 1980s, the US military has relied on a series of surveillance satellites referred to by the media as Keyhole satellites. Whatever their name, they are large expensive spacecraft, each weighing 10–15 tons and costing about $1.5 billion (of which a third is for the launcher). They orbit at between 300 and 900-km altitude and provide optical images that are said to have a resolution of 8–10 cm (3–4 inches) and less detailed infrared images. The four in orbit at the start of 2010 were launched in 1995, 1996, 2001 and 2005 and there must be some doubt about whether the two older ones are still fully functional. The latest two are upgraded versions that provide very wide coverage as well as detailed images. They can rotate their camera to dwell on a scene as they pass across the sky. The infrared feature means that they can take nighttime images. An additional satellite similar to them is under construction and due to be launched in 2010.

Four satellites leave big gaps in coverage. Typically, each flies over, say, Iran twice a day, spending 5–10 min each time with a view of some part of the country. Ten minutes × twice a day × four satellites is about an hour and a half, leaving the other 22½ h unwatched. And the orbits are reasonably predictable, giving plenty of time for things to happen between satellite visits and the evidence to then be concealed. That paranoid scenario is what underpins the multi-billion-dollar budgets spent on developing a scheme called Future Imaging Architecture – $10 billion is said to have been spent before it was cancelled in 2005. After 4 years of sucking its teeth, the military and intelligence eventually gave up on the dream of near continuous spying from space and decided to buy two very expensive satellites. One Senator who balked at the high price tag claimed that a single satellite would cost over $3 billion, although this figure has not been confirmed.

Perhaps referring to these two new spy satellites plus new radar imaging to be discussed shortly or eavesdropping satellites that are the subject of the next chapter, General Bruce Carlson (Figure 97), the Director of the National Reconnaissance Office that buys and operates them, said in April 2010 that "There are a number of very large and very critical reconnaissance satellites going to orbit in the next year, year and a half. We simply have to get these off and get them off on time". He described this as "the most aggressive launch campaign of the last 25 years".

General Carlson has been promoting a cost-saving concept called Next Generation Electro-Optical reconnaissance satellite, or NGEO. He says that it "will require a little bit of up-front investment" but will result in a modular and flexible satellite allowing NRO to insert new technology more rapidly than at present as well as reducing costs. Improving the technology to put into its expensive operational satellites is also addressed using tiny low-cost "CubeSats". NRO had 12 CubeSats under construction in 2009 according to NRO official Karyn Hayes-Ryan. She noted that they can be built in less than 6 months and thus enable NRO to "keep pace with Moore's Law", which we discussed in Chapter 5. Batteries, solar cells, computer processors, gyroscopes and radios were some of the technologies she listed as being tested on CubeSats. As well as quickly and cheaply testing new ideas in space, this approach helps train people and is "more risk tolerant", according to Ms Hayes-Ryan.

Figure 97. In 2009, retired USAF four-star General Bruce Carlson took over at the National Reconnaissance Office that buys and operates the USA's spy satellites. He is determined to halt the "erosion in the [NRO's] science and technology base" that he describes as "the seed corn for the future". Credit: NRO.

General Carlson notes that many of the current satellites have lasted far longer than expected and they deliver useful intelligence, mainly because of "the young people that write software" to adapt the images the satellites produce. He jokes that "we have satellites inside our very aging constellation that are old enough to vote and some, that are still operating, are old enough to drink. We don't let them drink, but they are old enough to drink". On another occasion, he quipped that "half the constellation is geriatric". He points out, however, that because of the improvements in software, "if you look at the product that we got ten years ago and compare it to the product that we have today, in many cases, it's an order of magnitude better in quality whether it's accuracy or clarity or timeliness".

A fifth satellite similar to the four just mentioned may also be in orbit at about 800-km altitude and is said to be a stealth satellite, impervious to detection from the ground. Launched in 1999, it was part of a program called MISTY that is thought to have to been cancelled in 2006 when projected costs got out of hand. Intelligence specialist Jeffrey Richelson, who first broke the story about the MISTY satellites, is skeptical about their stealthiness. He notes that civilian observers were able to keep track of the first MISTY satellite launched in 1990 and watch its various maneuvers.[2]

---

[2] Richelson (2008) pp. 179–180, Richelson (2002) p. 248, Covault (2006), Brinton (2009b), Brinton (2010c), Carlson (2009) pp. 5–6, Carlson (2010) pp. 2, 5, and Hayes-Ryan (2009) pp. 9–12.

Continuous monitoring of a region could, in principle, be achieved using geostationary satellites – 36,000 km above the earth and orbiting at the same rate as the earth beneath rotates, hence *geostationary*. The problem is the 36,000 km. Keyhole satellites are about 400 km above the earth, so 36,000 km is nearly 100 times further away and the resolution would need to improve accordingly. The current Keyhole satellites are thought to have resolution slightly better than turning the Hubble Space Telescope to point at the earth, which implies about 5–10 cm.[3]

The successor to the Hubble – the James Webb Space Telescope (JWST) – to be launched in 2014, will have resolution 2½ times better than Hubble, which is a long way short of the factor of 100 needed for a geostationary version of Keyhole. Even to achieve that 2½-times improvement, JWST has to unfold its mirror after reaching orbit because its 6½-m-wide mirror is too big to fit inside the rocket that will place it in orbit. Unfolding sounds fairly easy but the problem is that the unfolded mirror has to be smooth to an accuracy of one-fifty-thousandth of a millimeter, which is hard to achieve with a clunky opening mechanism. The 18 segments of which the JWST mirror is made have a small motor behind them to adjust them once in orbit.[4]

The technology for building even larger telescopes in space is a subject of some research. An unfolding concept that is 25 m wide has been analyzed, which would be 10 times the width of Hubble's mirror and give a resolution from geostationary orbit of about 50 cm to 1 m. But it would be extremely massive.

A somewhat more practical approach is to copy what is done in the biggest observatories here on earth. The famous Hale 200-inch (5-m) telescope on Mount Palomar was the world's largest (in the sense that its primary mirror was the largest) for 45 years until 1993, but is not even in the top 10 now. The biggest telescopes now are twice the size of the Hale but are made up of multiple mirrors carefully attached to each other to give a single mirror surface – continuous except at the joins of the segments. A massive 42-m mirror proposed for the so-called Extremely Large Telescope (ELT) will comprise 1,000 segments, each 1½ m across. JWST takes this approach to achieving its 6½-m-wide mirror – 18 segments, each $1\frac{1}{3}$ m wide. A variation on this idea is to have mirror segments with a few gaps or even a lot of gaps; an incomplete mirror with missing segments has some of the properties of a full mirror of the same scale – but lacks some others. The Very Large Telescope (VLT) owned by the European Southern Observatory on top of Cerro Paranal in Chile has just four segments, each of which is 65% bigger than the Hale 200-inch mirror. It is used to take ultra-detailed images of a very small area of the sky but is inappropriate for taking wide pictures.

A telescope in space that is a compromise between the full mirror of JWST and the very sparse VLT is shown in Figure 98. The concept is the work of Thales Alenia Space (France) and the French space agency CNES. It comprises nine mirror segments, each 1.8 m wide, which together would provide images from geostationary

---

[3]   Chaisson (1998) p. 29.
[4]   Mecham (2010).

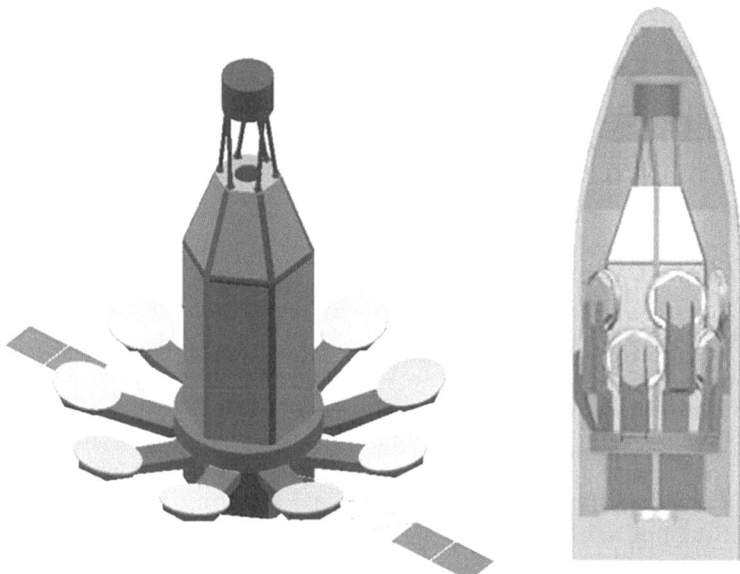

Figure 98. Design of a nine-segment deployable mirror for a geostationary observation satellite (left) and folded under the rocket fairing for launch (right). Credit: ESA, Thales Alenia Space/CNES.

orbit every second with a resolution of 1.2 m directly below and about 50 km on a side. Wider scenes would require multiple images of adjacent areas. A somewhat similar concept was studied by the US Air Force in the 1990s. Because it is a geostationary satellite, "directly below" means on the equator, which means the top of South America, the center of Africa, the tip of India and Sri Lanka, Malaysia, Indonesia and the Philippines. If your interests lie away from the equator, the resolution suffers, being about 5 m over southern Russia or Korea. Anything beyond 60° latitude will be very distorted.[5]

Ideas for even bigger space telescopes include shell structures that pop into the correct shape when freed from their rolled-up position – like a soft contact lens does. Membrane mirrors made of a thin film that can be unfolded in space have also been suggested. These concepts could lead to mirrors 100 m or more wide and thus provide resolution from geostationary orbit of about 10–25 cm, but achieving the required surface smoothness and robustness is very difficult.[6]

The images from a Keyhole satellite can be in several forms. Black and white is adequate for some purposes but there are times at which color helps to clarify what is happening. As with commercial satellites, Keyhole satellites can presumably take images in several colors ("spectral bands", to be more correct), which can be

---

[5] Mesrine (2008) pp. 25–26.
[6] Santer and Seffen (2009).

manipulated on a computer to highlight particular types of feature, such as recently disturbed land where a landmine or improvised explosive device (IED) might be buried. Infrared imagery is also available to see in the dark, although the detail is not as good as in daylight. Certain types of infrared imagery can also spot warm objects such as a car with its motor running or a building with central heating on. Stereo images can help clarify the scene in the image – many commercial satellites offer stereo as an option and Keyhole is presumably no different. The stereo clarifies many questions left unresolved by the mono images, such as turning lines on the image into ditches or fences.

The laws of physics make it possible to know when a particular satellite will next appear in the sky – and there are websites that will do the calculations for you.[7] If you think the USA is watching, you may be tempted to try to fool them. During the Cold War, the Soviets deployed decoys to try to fool the US satellites. Wooden aircraft, rubber ships, empty missile silos and more were used. Most were discovered as time went by. The lack of tracks following snowfall often showed when a site was a decoy. A rubber submarine broke in two following a storm.[8] Former CIA image analyst Dino Brugioni asserts that "a heavy snowfall negates all camouflaging efforts". He goes on to explain that the pattern of clearing the snow is a give-away, with the most important facilities cleared first, presumably the Headquarters, and others in descending order of importance – tracks to latrines are apparently high on the priority list. Melting snow on a roof suggests the building is occupied, while its absence suggests the opposite.[9] Brugioni also says that the best time to over fly a garrison is Sunday morning, when most of the equipment is at home – or Friday in an Islamic country, Saturday in Israel, etc.

A snapshot is interesting but a movie is better. In principle, a satellite moving across the sky can focus on a point on the ground and stay locked on that point for the few minutes that it is in view. The British Topsat satellite (mentioned again below) dwells on a scene for a few seconds for a different purpose, namely to get a single sharp picture of the scene, but that's still just a snapshot. The Russian Akron satellite mentioned below is thought to offer a "dwell" feature by swiveling its camera around to stay locked on the scene as the satellite passes overhead. This would allow you to see whether things or people are moving and in which direction. A small research satellite that allows ground controllers to control its video camera using a joystick has been tried out by the US military.[10] As noted above, the latest Keyhole satellites can dwell on a scene while it is in view – a matter of a few minutes at most.

Distinguishing between the green foliage of trees and the different green of the paint on a tank hidden underneath is beyond the ability of the Keyhole satellites.

---

[7]   For example, *www.heavens-above.com.*
[8]   Lindgren (2000) p. 142.
[9]   Quoted in Day *et al.* (1998) p. 219.
[10] Iannotta (2009).

The TacSat-3 satellite, launched in 2009, carries a sensor that will try to achieve just that. Perhaps even more importantly, it should detect freshly dug earth, which might help spot Improvised Explosive Devices, which currently cause 60% of US fatalities in Afghanistan. The technology is called hyper-spectral imagery (already mentioned in the previous chapter) and it breaks up the scene into hundreds of "colors". Satellites like Landsat (see Chapter 4) provide images with six or seven colors, some of which are in the infrared and so are colors that the human eye cannot perceive. The term "spectral bands" is more accurate than "colors", so these satellites are said to provide multi-spectral images. NASA in 2000 and the European Space Agency in 2001 put sensors in orbit that could distinguish more than 200 different spectral bands for scientific purposes, which is so many more than Landsat's half dozen that a new term was invented to describe it – "hyper-spectral". TacSat-3 is the first acknowledged launch of this technology by the US military. By choosing the right combination of bands (i.e., colors), you can distinguish between otherwise similar features on the ground. For example, hyper-spectral images should allow the type of paint or surface coating on a missile, aircraft, ship, tank, etc. to be analyzed.

TacSat-3 goes a step further by having in its computer a menu of colors of different objects and surface types – a kind of signature for each. The satellite not only takes a hyper-spectral image of the scene below, but it also checks against its store of signatures for anything of interest and relays that information to the troops on the ground below. This is the "tactical" part of the TacSat name. If the technique is successful, the Air Force plans to launch a fleet of these small satellites (TacSat-3 cost $85 million, of which $25 million was for the launcher).[11]

The four or five Keyhole optical satellites are complemented by a series of radar imaging satellites, about which more shortly, and by the use of imagery purchased from commercial satellites, which we will now discuss.

COMMERCIAL IMAGING SATELLITES

The detail available in space images such as those on the front cover of this book is discussed in Chapter 4 (commercial). They were provided by GeoEye, which is one of two US companies offering images from space with a resolution better than 1 m. For the first 30 years of the space age, images such as these were only available to the intelligence community in the USA, the Soviet Union (as it then was) and a few of their close allies. Ironically, the highly secretive communist world was the first to challenge this situation. In 1992, 3 years after the fall of the Berlin Wall, Russia began offering images for sale from its hitherto secret spy satellites. The images were of about 2-m resolution and were copied from the conventional film that was sent back to earth in a recovery capsule from the short-lived satellite – short-lived because once the film was used up, the satellite had no further value. That imagery is still

---

[11]   Matthews (2009).

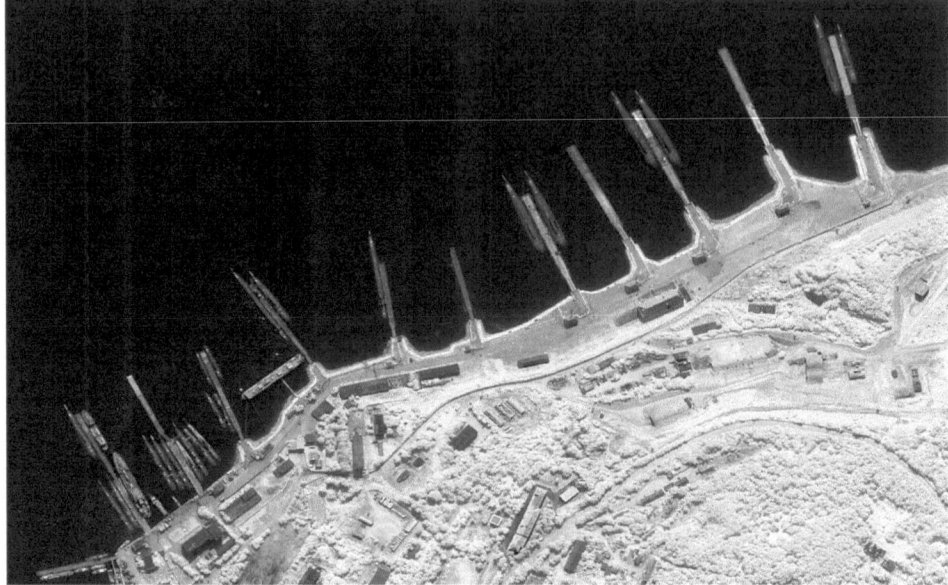

Figure 99. Russian submarine base in Kamchatka on the Pacific Coast. © 2010 ImageSat International N.V., licensed by ImageSat International N.V.

Figure 100. Ramenskoye Flight Test Centre near Moscow. © 2010 ImageSat International N.V., licensed by ImageSat International N.V.

Figure 101. Russian naval base near Odessa, leased from Ukraine. © 2010 ImageSat International N.V., licensed by ImageSat International N.V.

available on request (but not of Russia), although the most recent of the Resurs series of satellites, Resurs-DK, launched in 2006, has done away with the recoverable-film approach and transmits digital images to ground by radio link.

Responding to requests from US industry to be allowed to launch satellites with similar resolution, the US Administration reacted in 1994 by permitting the sale by US companies to all but a few countries of imagery with a resolution of 1 m. This limit on resolution is not spelled out in the 1994 Presidential Decision Directive no. 23; instead, the Department of Commerce decides on a case-by-case basis whether to issue a license for a specific satellite system. Thus, in 2000, a license issued to EarthWatch permitted that company to acquire data with 50-cm resolution, so that company then lowered the orbit of its existing QuickBird-1 satellite to improve its resolution from 1 m to 61 cm. There are rules about holding back imagery for 24 h so that military maneuvers are not compromised and about not selling imagery of better than 60-cm resolution, and the US Government can introduce other ad-hoc constraints on the imagery when necessary.

US industry got a boost in 2003 when the National Geospatial-Intelligence Agency (NGA) placed two contracts guaranteeing to purchase $150 million per annum of imagery from each of two suppliers and offering about half of the $500 million cost of building and launching each of the high-resolution optical satellites.[12] The action by NGA followed on from a $100 million contract for high-resolution

---

[12]   de Selding (2009b).

imagery in 1999 to a third company. A smaller contract followed in 2004 to that third company and in 2005, that company merged with one of the 2003 winners, leaving DigitalGlobe and GeoEye as NGA's two high-resolution commercial suppliers as this book goes to press.

The NGA is the arm of the military and intelligence communities that "develops imagery and map-based intelligence solutions for U. S. national defense, homeland security and safety of navigation".[13] In other words, they process the imagery from spy satellites. The decision by NGA to buy high-resolution imagery from commercial companies was a major change of policy, since it provides a glimpse into the otherwise highly secretive world of spy satellites and intelligence agencies. When NGA orders some images from either GeoEye or Digital Globe, the location to be imaged has to be identified and this information is sensitive. It's not much of a surprise if NGA wants images of Afghanistan, Iran or North Korea. But what if they want images of, say, Mexico or France or any other US ally? The commercial companies have set up military-grade security facilities to deal with this issue, so that only specifically authorized staff deal with the military and information about where and when images are being taken is encrypted and guarded.

NGA thought the imagery was sufficiently useful that it bought up all available imagery of Afghanistan in the period after 9/11, thus denying those images to anyone else. This method for ensuring the adversary doesn't get access to satellite images is popularly known as "buy to deny" or "checkbook shutter control".

The commercial GeoEye and DigitalGlobe imagery is an important addition to the military's own surveillance satellites because of the heavy intelligence requirements of the Afghanistan and Iraq campaigns. The military appetite for satellite imagery is insatiable during a major conflict, resulting in the need to prioritize who gets access to the images. The Army may want images of a region in which they are involved in military operations, whereas the State Department or intelligence agencies may want regular imagery of Iran's nuclear facilities. One well known consequence of the limited amount of satellite imagery was the failure to adequately monitor the Afghanistan/Pakistan border as the Taliban and Al Qaeda forces escaped from the USA-backed Northern Alliance forces in 2001 and 2002. Instead, satellite imagery was increasingly focused on Iraq, starting with the run-up to the invasion in 2003 and continuing thereafter. One US General in Afghanistan said in 2006 that the US military's "biggest mistake" was the lack of satellite surveillance in the south of that country because the Iraq War had consumed it all.[14] The quality of imagery from this type of satellite relevant to the military services is illustrated in Figures 99, 100 and 101.

The US military and intelligence community has long been the largest customer for France's SPOT satellite images and that relationship continues, currently running at about $5 million per year.[15] SPOT images are not as detailed as those of

---

[13]   NGA website, *www.nga.mil.*
[14]   Rashid (2009) pp. LVII, 134, 223, 357, 359.
[15]   *Space News* (2010c).

Figure 102. Vice Adm. Robert Murrett, Director of the National Geospatial-Intelligence Agency, downplayed the impact on commercial satellite owners of a decision to buy two government-owned imaging satellites: "We are the single strongest supporter of the commercial remote sensing industry. It is absolutely integral to our success and is a fundamental building block for what we do as an agency," he said.[16] Credit: NGA.

GeoEye or DigitalGlobe but give a wider perspective, which is useful for planning a military mission. More generally, the market for commercial satellite data in 2009 was $1.2 billiion, of which the US military represented a third, the military in other countries a quarter, and the civilian sector comprising less than 40%.[17]

The NGA is sufficiently content with the GeoEye and DigitalGlobe arrangements to want to extend them, but this time, it wants the private sector to finance the purchase of two new satellites rather than the government funding half as for the current satellites. In the same breath, the National Reconnaissance Office (NRO) has been authorized to buy two next-generation spy satellites said to be astronomically expensive.[18] It remains to be seen how this tentative acceptance of commercial satellites by the intelligence community will evolve (Figure 102).

RUSSIAN IMAGING SATELLITES

The Satellite Intelligence Directorate within Russia's military intelligence organization, the GRU, seems to be responsible for most spy satellites – of both the imaging and eavesdropping (Chapter 9) kind. A "space branch" of the armed forces looks after the launch of the satellites into orbit and tracking them in space and as of 2001, is independent of the individual services. Some friction has surfaced from time to

---

[16] Shalal-Esa (2009).
[17] Keith (2010a).
[18] Brinton (2009b), Brinton (2010d), and de Selding (2010e).

time between the Navy and the other services (see Chapter 9 on eavesdropping satellites) but, by and large, the Soviet Union and now Russia has had a fairly coherent military space program. The civilian intelligence agency, the KGB and its successors, seems not to be involved. Thus, the management scheme in Russia appears to be simpler than in the USA, where the CIA is very much involved and often at odds with the needs of the military services – the CIA being often interested in strategic or long-term topics such as nuclear weapons or locating Al Qaeda leaders while the Army, for example, wants information about today's battlefield. The inter-service bickering of the 1960s seems to have been reduced by the creation of the National Reconnaissance Office (NRO), which takes responsibility for most imaging satellites – in the past, the Air Force and the Navy often had separate systems, which now is only the case for communications satellites.[19] The National Security Agency (NSA) is responsible for most of the USA's eavesdropping satellites, which is another forum in which CIA and US military may have conflicting priorities. The Russian Satellite Intelligence Directorate seems to be roughly equivalent to America's NRO plus some of NSA.[20]

Russia has been operating imaging satellites since the 1960s, having started like the USA with wet-film cameras that had to be returned to earth to have the film developed before the images could be seen and analyzed. The Zenit-2 series, as it was known, has been widely publicized and, like America's CORONA series, was instrumental in enabling the SALT-I and SALT-II strategic arms-limitation treaties to be negotiated between the USA and the Soviet Union. These landmark treaties brought to an end the hitherto unstoppable increase in the nuclear arsenals of the super-powers, and sharply reduced the likelihood of nuclear war.[21]

While the USA transitioned to digital camera technology and radio transmission of the images to ground in the 1980s, Russia was more than 10 years behind. Indeed, wet film is still used in Russia's only operational spy satellites, which fall into two general categories. The 10½-ton Yenissey (sometimes called Orlets-2) satellite carries a staggering 22 recoverable capsules, chucking them out when a chunk of film is ready to be analyzed until after about a year, the whole camera and the remaining film is returned in a large capsule and can be refurbished to be used again. Only two have been launched, the most recent in 2000. An apparent close cousin, Don (Orlets-1), is said to weigh 4 tons less, at 6½ tons, and carry eight capsules. One is thought to have been launched in September 2006 but returned to earth 2 months later, and it may be that this class of satellite has been discontinued.

The second series of satellites still using wet film is the Kobalt-M (sometimes called Yantar-4K), which carries two small (¼-ton) and one large return capsules.

---

[19]   When I began work on space systems in 1996, I was analyzing data from separate Navy, Army and Air Force satellite systems, helping John Berbert of NASA to compare their accuracy in measuring earth's gravity field (not to mention three of NASA's own systems plus those of various US scientific groups).

[20]   Hendrickx (2005) p. 99.

[21]   As told in my previous book, *Spies in the Sky*, Norris (2007).

One of these is currently launched every year or so, each lasting about 3 months before returning. The most recent of this type was launched in April 2010, with previous ones in 2004 (3½ months in orbit), 2006 (2½ months), 2007 (2½ months), 2008 (3½ months) and 2009 (3 months). Their orbit is slightly elliptical, with low and high points of about 170 and 340 km. The satellite is maneuvered so that the low point is over the area of greatest interest, where the resolution of the images will be best (rumored to be a few centimeters), and the elliptical shape of the orbit avoids unnecessary atmospheric drag, which would cause the satellite to re-enter the atmosphere sooner than planned. They weigh 6½ tons and continue the Russian tradition of putting the camera and film in a pressurized cabin. The camera and telescope are returned in the final large capsule so that they can be flown again on a future mission. From the beginning of the space age, the USA allowed the vacuum of outer space to penetrate its robotic satellites, thus minimizing weight but requiring special electronics, materials, grease, film, mechanical mechanisms and so on to cope with the vacuum – many materials, such as plastics and foam, give off gases in vacuum, which contaminates lenses and mirrors in the camera, while mechanisms freeze because normal lubricants evaporate. The Soviet Union and now Russia decided to keep normal atmospheric pressure inside many robotic satellites, thus avoiding the need to develop vacuum-compatible materials, films, mechanisms, etc. The penalty of pressurization is heavy weight because of the need to seal in the air – and that means big launchers with big price tags. Modern Russian satellites often still choose the pressurized approach – for example, all of the Russian Glonass navigation satellites (see Chapter 6) launched to date have been pressurized, although the first of a new version open to the vacuum of space, Glonass-K, is due to be launched in 2010.

A special variant of the Kobalt-M called Kometa (Yantar-1KFT) is launched every few years to take high-quality survey-type images for updating maps – this variant returns the whole film in one go after about 6 weeks without any intermediate capsules.

Three types of Russian spy satellite send images back by radio link, although all seem to be either defunct or out of order as this book is being written. About 10 of the Neman series (confusingly also called Yantar-4KS) were launched up until 2003 and radioed back their images either direct to ground or via a relay satellite (see "Getting the images back quickly" below). Two of the Akron or Araks satellites have been launched but the last one (launched in 2002) seems to be inoperative. They were unusual in that their orbit was more than 1,500 km high, thus limiting the resolution of the images. However, they may have been able to swivel the camera as they passed over a scene and dwell on it for a few minutes – giving a short video capability a bit like that on most of today's digital cameras. Like Neman, images were radioed back to earth either directly or via relay satellites.

The most recent digital-imaging Russian satellite was well publicized when launched in July 2008, but seems to have failed soon thereafter. Called Persona, it was derived from the civilian surveillance Resurs-DK satellites and was placed in a 700-km circular orbit. The idea was to merge the top-quality camera from the high-flying Arkon/Araks with the superior onboard storage and computing ability of the civilian Resurs-DK. The lower orbit should have improved the image resolution to about 30 cm.

All in all, then, the Russian military have not had great success in recent years with their digital camera satellites and are still relying heavily on wet-film cameras that have to be physically returned to earth to be developed before being analyzed.[22]

Russia's commercial imaging satellites have made a more successful transition to digital technology. As described in Chapter 4, Russia initiated the sale of imagery with resolution of better than 2 m in the early 1990s by selling slightly blurred versions of its military satellite images. The USA then legalized the sale of such images and a whole range of companies and countries are now selling them. Those early Russian images came from traditional film and were scanned into a computer for commercial sale.

Russia still sells that type of image, but in addition, in 2006, launched the Resurs-DK, a civilian version of the Neman military satellite mentioned above and a more modern version of the Resurs-01 medium-resolution series.[23] Like many Western satellites, the Resurs-DK offers images with 90-cm resolution in black and white or 1 ½-m in three color bands. To achieve this resolution, the satellite flies in a slightly elliptical orbit – 360 km above the earth at point of closest approach, rising to 600 km at the highest point. The best resolution is achieved when the satellite is at the lowest part of the orbit. It may take some time and require the burning up of some of its fuel to swing the lowest point of the orbit to where a customer wants a 90-cm-resolution image. Weighing 6 ½ tons, Resurs-DK is expensive to build and launch compared with many Western satellites that provide comparable images. It requires quite a lot of fuel to power its rocket motors to alter its orbit in response to requests for particular images. It may also contain a heavy and bulky pressurized cabin in which to place its electronics and other critical parts and thus protect them from the vacuum of space (see above).

An even more modern version called Resurs-P is due to be launched in late 2010 or 2011, with cameras that are expected to be identical to those on the existing civilian Resurs-DK.

In a departure from the norm, Russia is buying satellite technology from the West for a new commercial satellite series called Kanopus-V. The cameras and radio systems are Russian but the satellite technology comes from Britain's Surrey Satellite (see Chapter 5). The heavy Cold War heritage of Resurs-DK contrasts sharply with the modern Kanopus-V, which weighs in at just 400 kg, but their image quality is similar – Kanopus-V promises black-and-white images with 2-m resolution. The first in the series is due to be launched in 2010.[24]

---

[22] Covault (2008a), Lardier (2010c), *Air & Cosmos* (2007b), Lardier (2008c), Morring (2007a), *Air & Cosmos* (2010), and Lardier (2010b).

[23] Easily confused with the military Resurs-F1 and Resurs-F2 series of wet-film imaging satellites.

[24] Lardier (2010c), Lardier (2009c), and Grishin (2009).

Figure 103. US Air Force Chief of Staff General T. Michael Moseley said "killing another nation's satellite is an act of war". He also notes the vulnerability of satellites in geostationary orbit: "... if you can hit something at 500-plus miles in orbit, you can certainly hit something out beyond 20,000 miles."[25] Credit: US Air Force.

OTHER COUNTRIES

In this section, we take a quick look at the military imaging satellites of China, Europe, Israel, India and Japan.

China

China's space activities have had a dramatic impact recently, but because of shooting satellites down as much as putting them up. In January 2007, China destroyed one of its defunct weather satellites 850 km out in space by firing a missile at it from the ground. The event was presumably intended as a shot across the bows of the USA and, if so, it succeeded. Three months after the event and having had time to assess its consequences, US Air Force Chief of Staff General T. Michael Moseley (Figure 103) called it a "strategically dislocating event that is no different than when the Russians put Sputnik up in October 1957". About 6 months earlier, China beamed a laser from the ground onto a US reconnaissance satellite. It did no harm "but it

---

[25]   Fulghum and Butler (2007).

made us think", said Donald Kerr, the Director of the National Reconnaissance Office that operates US spy satellites.[26]

Having increased the amount of debris in space by 10–20% by destroying its old satellite, China began an initiative about the same time to place a fleet of surveillance satellites in orbit – which will be just as much at risk of being hit by a piece of high-speed debris as anyone else's satellites. Between April 2006 and March 2010, nine satellites in the Yaogan series have been launched – five[27] of them are probably 2-ton imaging radar satellites and the others are somewhat smaller and carry optical cameras. The first seven are in orbits about 500–650 km high, while the two most recent are about 1,100–1,200 km high. The most recent (March 2010) is different from the first eight in staying below 63° latitude, while the others go within 8° of the North and South Poles. One or both of the two most recent Yaogans may therefore be electronic ocean surveillance (eavesdropping) satellites like the US NOSS and Russian EORSAT satellites (see Chapter 9), although some commentators seem content that they are the latest in the optical and imaging radar series.[28] The higher orbit of the two most recent might even be a recognition by the Chinese that their destruction of a satellite in 2007 made orbits below 850 km prone to collision with debris – augmented by the accidental collision of an American Iridium communications satellite with a defunct Russian Cosmos satellite in 2009 at 800-km altitude.

Officially, the purpose of the Yaogan satellites is to study the earth's resources, forecast crop yields, map the consequences of natural disasters and the like, but most commentators consider them to be essentially military in nature. The radar satellites produce images even at night and through clouds, while the optical satellites provide color and other details that help characterize objects and fields in the scene. They all radio the images back to earth and thus avoid the need for re-entry capsules. It's not clear whom China considers its main adversary to be, but the deployment of these surveillance satellites is consistent with the super-power ambitions illustrated by its other military investments, including missiles (see "Missile early warning", below).

France and continental Europe

France was the first European nation to fly civilian surveillance satellites – the SPOT series, described in Chapter 4. During the 1991 Gulf War, France was part of the

---

[26] Covault (2007d), and Fulghum and Butler (2007).

[27] Some commentators identify six as being radar imaging – see, e.g., *www.nasaspace-flight.com/2009/12/china-completes-2009-schedule-by-launching-another-spy-satellite/*; five of them were launched on the Long March 4 rocket, but one, Yaogan-6, was launched in April 2009 on the smaller Long March 2 rocket, as were the optical satellites, suggesting that it, too, was of the optical type.

[28] Lardier (2010a) favors the imaging satellite option, while McDowell (2010) states that Yaogan-9 is accompanied by two sub-satellites and is thus likely to be an ocean surveillance satellite.

USA-led allied force that pushed Iraq out of Kuwait, but the Head of State, President Mitterand, felt that it didn't have adequate access to the best American spy satellite images during the military campaign. Immediately after that conflict, France began a program to place high-resolution spy satellites in orbit, culminating with the launch of Helios-1A in 1995. This was followed by the identical Helios-1B in 1999 and two second-generation satellites, Helios-2A and -2B, in 2004 and 2009. The Helios-2 satellites weigh over 4 tons each and provide 35-cm-resolution optical images. They also take infrared images at night with resolution about twice that of the daytime images. In addition to the high-resolution camera, Helios-2 has a wide-area camera to provide a survey and planning capability. The cost of the two second-generation satellites, their launchers, the ground facilities to receive and process the data and 10 years of operation came to about $2.9 billion, of which about half is the cost of the satellites. At the start of 2010, Helios-1A, Helios-2A and Helios-2B were all functioning. In late 2010, the first of a pair of 1-ton satellites called Pléiades will be launched. Costing about $200 million each plus nearly the same again for launchers and ground facilities, Pléiades will offer 70-cm resolution and the satellites are intended to be shared between military and commercial users.[29]

Rather than build imaging radar satellites that can see through cloud, France has entered into an agreement with Germany and Italy to have access to data from their satellites (see "Radar imaging", below) in return for sharing Helios images with them. Belgium, Greece and Spain also get access to some Helios images in return for a cash contribution to the program budget. These countries are now looking at merging their activities for the next round of satellites within a program called MUSIS, and thus avoid duplicating the investment, and Poland and Sweden have stated their intention to join in. Despite the obvious economic sense in avoiding duplication, the six original countries are finding it difficult to agree what each one *won't* do – for example, which of Italy or Germany will give up its imaging radar activity. This argument will get even more complicated when Spain launches its Paz imaging radar satellite in 2012 – primarily a military system, although similar in performance to the German civilian TerraSAR-X.[30] The usual European solution to the question of who stops making something is for all of the countries to continue doing it, thus weakening the economic argument for collaboration. Occasionally, a real compromise is achieved, such as by having the relevant industry in all the countries merge. Faced with competition in the global marketplace, industry can usually make the hard choices, while the politicians rarely can. The downside of merging companies for this kind of reason is that the merged entity is likely to be something of a sluggish monster, relying on state handouts for years to come.

Europe has developed the Ariane rocket in order to ensure that it can launch its own satellites when it wants to (Figure 104). In 1972, before Ariane existed, the USA

---

[29] Lardier (2008b), Bombeau (2009), de Selding (2010d), de Selding (2010f), and Taverna (2010).
[30] de Selding (2009c), and *Space News* (2010b).

Figure 104. Europe's Ariane rocket, seen here launching France's Helios-2B spy satellite, and its launch site in French Guiana were built to avoid dependence on US launch vehicles. Credit: Arianespace/CNES/ESA.

agreed to launch the European Symphonie satellite, provided it wasn't used commercially. This US precondition for launching Symphonie was perceived in Europe as the USA protecting its satellite industry. The 1970s were the height of the Cold War when the only other launch supplier, Russia, posed Europe's main military threat and so was not a viable alternative to the USA. Europe's response was to focus on developing its own rocket, resulting in the first flight of Ariane in 1979. European suspicion about US motives was further fuelled by a similar "no commercial use" restriction imposed by the USA on the German MOMS/SPAS camera in 1983.

Now that Ariane is available when needed, some European countries are happy to use other launchers if they are better suited to the task. The German SAR-Lupe satellites used Russia's Cosmos rocket, while Italy's COSMO/SkyMed used the US Delta. Since France is the biggest investor in Ariane, the French Helios satellites have stayed loyal to that launcher (Figure 104) – but Pléiades will be launched on a Russian rocket in the interests of economy.

Great Britain

Britain has relied on US spy satellites because of the agreement to share intelligence information dating back to World War II and the collaboration to develop the

Figure 105. Air Chief Marshall Sir Stephen Dalton, professional Head of Britain's Royal Air Force, says "resource constraints may mean that we'll have to continue to rely on alliances and partnerships for access to space. However the extent to which the UK relies almost entirely on third party capabilities is a potential cause for concern. Arguably, we should cast the net more widely in looking for partners".[31] Credit: Sgt Andy Malthouse ABIPP, Crown Copyright/MOD 2009.

atomic bomb. The Iraq and Afghanistan conflicts since 2001 have illustrated the limitations of this arrangement, namely that if the USA hasn't got enough spy satellites for its own forces, the UK is going to find it hard to get the space-based intelligence it needs. A prototype military imaging satellite called Topsat was launched in 2005, providing images with a resolution of 2½ m in black and white and 5 m in color. Recent UK Government statements have confirmed Britain's unease at the reliance on the USA. The Defence Minister recently told Parliament that "we might increase our contribution to allied space capabilities or invest in our own national capabilities" (Figure 105). This suggests that Britain might contribute directly to the US military space activities – as happened once before in the eavesdropping arena (see next chapter). Or it might mean that Britain will join with the European MUSIS consortium.[32]

For the moment, Britain remains the only Permanent Member of the United Nations Security Council and the only nuclear power recognized in the Non-Proliferation Treaty that lacks its own operational military space surveillance system.

---

[31] Dalton (2010).
[32] Secretary of State for Defence (2010), and Dalton (2010).

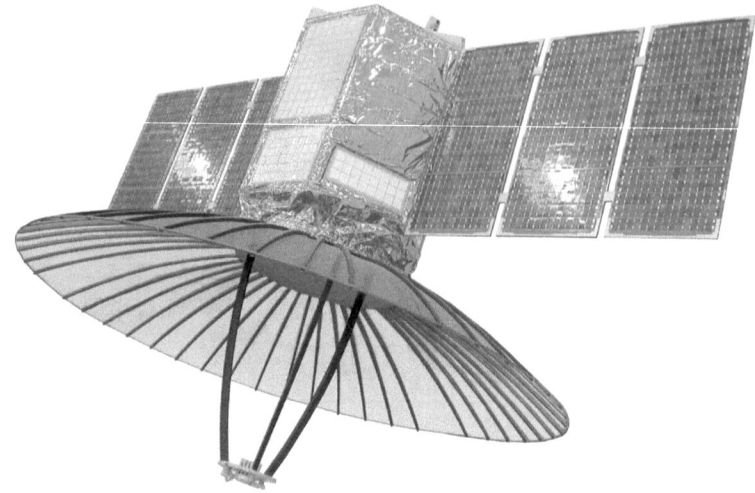

Figure 106. Model of Israel's TecSAR imaging radar satellite, launched in 2008. Credit: Israel Aerospace Industries (IAI).

Israel

Despite its small size, Israel has an impressive space program, including manufacturing spy satellites and the rockets with which to launch them into orbit. Three of its military optical satellites are currently operational: Ofeq-5, -7 and -9. Israel also has the TecSAR imaging radar satellite (Figure 106) that was placed in orbit by an Indian rocket in 2008. In addition, Israel military have access to the images from the civilian EROS A and EROS B1 satellites that were built in Israel and are operated by the commercial ImageSat company. The civilian EROS B satellite has a resolution of 70 cm (see Figures 99, 100 and 101) and, presumably, the Ofeq series is better than that. TecSAR images have a best resolution of about 1 m, but have the great attraction that they see through cloud and at night. All are small by comparison with US and Russian spy satellites, weighing in at about 300 kg each. They orbit about 500 km high. Israeli intelligence analysts therefore have access to seven military or quasi-military surveillance satellites, giving them the ability to get imagery of an area of interest within a few hours if the weather is clear – 2 or 3 days if not (using TecSAR).[33]

A peculiarity of the satellites that Israel launches itself is that they go round the earth "the wrong way", from right to left rather than the left-to-right direction of almost all other satellites (looking at them from, say, the moon with the earth's north pole at the top). This policy is not an idiosyncratic extension of the right-to-left

---

[33]   Opall-Rome (2008), and Morring (2010b).

direction of Hebrew script, but is dictated by the need to launch westward over the Mediterranean Sea to avoid empty rocket casings falling on countries to the east. Unfortunately for Israel, the earth rotates from left to right, so the rocket has to counter that movement as well as reaching the 18,000 mph (29,000 kph) needed to get into orbit. The result is that Israel's rockets carry about 30% less mass into orbit than they could if they launched in the other direction.[34] This unhappy effect of geography explains why Israel is keen to build a relationship with India, whose launch options are not constrained in that way.

India

India has been associated with Israel's military space endeavors by virtue of launching Israel's TecSAR imaging radar satellite into orbit on its PSLV rocket in 2008 and using an Israeli radar on India's Risat-2 satellite, launched in 2009. This collaboration is consistent with the emergence of Israel as India's largest supplier of defense equipment after Russia. The range of equipment with which Israel is assisting India includes air defense, submarine-launched nuclear missiles, hypersonic engines, cruise missiles and electronic warfare systems. A follow-on imaging radar satellite called Risat-1 containing more Indian and less Israeli technology is due to be launched in 2010.

Although India claims not to have specifically military satellites, some of those launched recently are of particular use to India's military. Examples include the Cartosat-2A satellite, with 80-cm resolution, launched in 2008, and Risat-2, already mentioned. Cartosat-2B, launched in July 2010, provides a level of redundancy needed for an operational system. The Indian military have made no secret of their interest in space (see Figure 107) and there has recently been evidence of closer ties between the civilian space agency, ISRO, and the Indian Army.[35]

Japan

As a non-nuclear power, Japan had relied on US strategic facilities until the North Koreans began testing missiles in the mid 1990s that could span the Sea of Japan separating the two countries – 600 km at the nearest point. Up until that point, Japan had avoided any military involvement in space but North Korea's missile tests in 1998 and nuclear bomb ambitions changed that. With considerable capability in scientific and commercial satellites built up over the preceding 20 years, Japan began a crash program to deploy both optical and radar military surveillance satellites. The

---

[34] Dr Zvi Kaplan (Israel Space Agency), presentation to the Royal Aeronautical Society (London), June 17th 2010.

[35] Singh (2008), and Jayaraman (2008).

Figure 107. Indian Army Chief of Staff Gen. Deepak Kapoor said "India needs to optimize the use of space for military applications to counter China's rapid strides in the sphere". Credit: Indian Army.[36]

first Information Gathering Satellite (IGS) was launched in 2003. The intention is to have two optical and two radar imaging satellites in orbit at all times, but this objective has proved elusive due to a rocket failure in 2003 that destroyed the third and fourth satellites and early failure of the first radar satellite in 2007. Three more launches in 2006, 2007 and 2009 lofted a total of five more satellites into orbit, although some have experienced problems. The two most recent optical satellites are of a more advanced design and have a resolution of about 60 cm, while staying in the small satellite category – each weighs about 850 kg. The sole operational imaging radar satellite weighs about 1.2 tons and builds on technology that was developed by Japan's civilian space agency, JAXA, for scientific observation of the earth – its image resolution is about 1–3 m. The most recent in the series of optical satellites, IGS-3a, was launched in late 2009, costing over $500 million. As of early 2010, two optical and one radar imaging satellite are operational. Further launches have been announced for optical satellites in 2011 and 2014 and of radar imaging satellites in 2011 and 2012.

The Initial IGS program cost $2.2 billion and the yearly cost is said to be about $350 million. The cost of each new satellite is said to be about $450–550 million but it is not clear whether this is in addition to the $350 million.[37]

---

[36]   Jayaraman (2008).
[37]   Morring (2007b), *Space News* (2007), Kallender-Umezu (2009), Morring (2009), and Lardier (2010c).

Figure 108. Japanese Defense Minister Yasukazu Hamada, left, seen here in Singapore in May 2009 with his US and South Korean counterparts, Robert M. Gates, center, and Lee Sang-Hee. A month earlier, he said in Japan's Diet (Parliament) that Japan should consider developing an early warning satellite to monitor launches across its territory – North Korea had launched a rocket over Japanese territory on April 5th. Credit: DoD photo by US Air Force Master Sgt Jerry Morrison.

The continued provocation of North Korean rockets and missiles being fired across the Sea of Japan[38] towards the Japanese homeland has brought calls for additional military space programs. The Japanese Minister of Defense (Figure 108) recently spoke in favor of a missile early warning satellite to monitor North Korea's actions – see the section on "Missile early warning" below for more about this type of satellite. However, civil servants in the Ministry of Defense cautioned that financial constraints made any such new program unlikely and hinted that the Minister's remarks were primarily a political response to the North Korean rocket firing 4 days earlier.[39]

---

[38] One of the few subjects on which North and South Korea agree is that the Sea of Japan should be called the East Sea.

[39] Kallender-Umezu (2009).

RADAR IMAGING

In the summer of 2008, the US Administration declassified the fact that the country has radar imaging satellites. This was hardly earth-shattering news, since the "secret" had been widely known and discussed for more than a decade. It is also common knowledge that US efforts to develop a new generation of imaging radar satellites had become so expensive that the effort was cancelled in 2008.[40] There are thought to be four of the current generation of radar imaging satellites in orbit, launched in 1997, 1999, 2000 and 2005, which means that a replacement for the longest-serving one is long overdue. Called "Lacrosse"[41] by the media, they are extremely large (50 m across with their solar arrays deployed) in order to generate the 10–20 kW of power needed to drive the radar – and correspondingly expensive, at $1½ billion each. They provide imagery through cloud and in the dark with a resolution of about 1 m.[42]

Not satisfied with the infrequent radar images that the four Lacrosse satellites provide, the USA aspired to build a fleet of satellites that would provide nearly continuous coverage of areas of interest. The initiative was called Space Radar, involving up to 20 satellites and costing more than $25 billion in some scenarios. In 2008, the plan was scrapped and a decision taken to purchase radar imagery from military and commercial non-US systems, such as those operated by Germany, Canada, Italy and Israel. In addition, one or two satellites to demonstrate a next-generation imaging radar are still being developed that might lead to a replacement for Lacrosse, with first launch forecast for 2012. Thus, in the last few years, the USA has backed away from two very expensive spy satellite concepts – the Future Imaging Architecture mentioned above and Space Radar. Perhaps this more cost-conscious approach explains why the number two official at the National Reconnaissance Office (NRO), Betty Sapp, was able to tell a Congressional Subcommittee in April 2010 that "the NRO received an Unqualified Opinion on its fiscal year 2009 Financial Statement". More surprisingly, she added that "this was the first clean audit for a defense intelligence agency since 2003".[43]

This sequence of events began in mid 2008 when the USA began to buy small quantities of radar imagery from Canadian company MDA, which owns and operates the Radarsat satellites. Then, in early 2010, the NGA placed contracts to buy up to $85 million-worth of data over a 5-year period from each of three foreign imaging radar satellite families: Germany's TerraSAR-X and Tandem-X, Italy's four COSMO/SkyMed satellites[44] and Canada's Radarsat-1 and -2. The USA is also

---

[40]   Brinton (2009d).
[41]   ONYX is the name used by one authoritative commentator who also claims there are only three in orbit – he excludes the 1999 launch; Richelson (2008) p. 180.
[42]   Covault (2006).
[43]   Sapp (2010) p. 8.
[44]   The fourth is due to be launched in 2010.

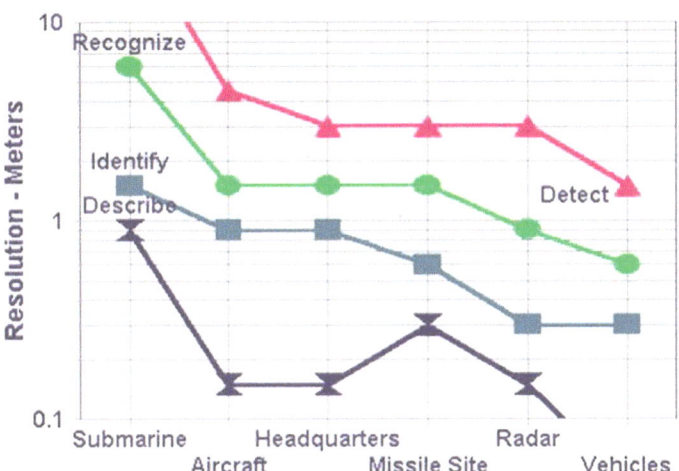

Figure 109. Illustration of how image resolution translates into military information.[45]

said to be considering buying a small imaging radar satellite based on the Israeli TecSAR technology (Figure 106). US Southern Command has already purchased some TecSAR imagery, probably to help in the war on drugs in Central and South America.[46]

As Table 5 shows, several countries now have imaging radar satellites in orbit with sufficient resolution to be of interest to the military.

Figure 109 illustrates the utility of better image resolution to the military, distinguishing between *detecting* that something is there, *recognizing* that it's a tank or an aircraft, say, *identifying* its specific type such as a T-54 tank or MIG-21J aircraft, and *describing* or technically analyzing it in detail. It is a very rough guide because lots of factors other than resolution influence the information in an image such as whether the object of interest is camouflaged in any way, the level of contrast in the image, the brightness of the object compared to other objects, glare and glint, differences in texture between the object and its neighbors, etc. The figure indicates that an image with 1-m resolution (horizontal line across the middle) allows you, for example, to identify a submarine but not describe it in detail, to recognize and just about identify an aircraft, and to detect but not identify a vehicle. An image with 3-m resolution (the second horizontal line above the 1-m line) allows you, for example, to detect but not identify missile sites and radars. Lower in the figure, an image with 60-cm resolution (four horizontal lines below the 1-m line) allows a missile site to be identified and a vehicle to be recognized.

---

[45] Adapted from NATO STANAG 3769 Annex C "Minimum Resolved Object Sizes for Imagery Interpretation".
[46] Brinton (2010a), Doyle (2008), Brinton (2009d), and Butler and Taverna (2010).

Table 5. Current (July 2010) high-resolution imaging radar satellites.

	Country	Best resolution	Frequency band, wavelength	Weight (tons)	Cost $M
Radarsat-1/-2	Canada	3 m	C, 5 ½ cm	2.2	1,500
TerraSAR/ Tandem-X	Germany	1 m	X, 3 cm	1.3	430
SAR-Lupe- 1/2/3/4/5*	Germany	50 cm	X, 3 cm	0.8	400
COSMO/SkyMed- 1/2/3	Italy	< 80 cm	X, 3 cm	1.7	1,400
Envisat	Europe	30 m	C, 5 ½ cm	8.2	3,000
TecSAR*	Israel	< 1 m	X, 3 cm	0.3	~ 150
IGS-2B*	Japan	1–3 m?	X, 3 cm	1.2	500
Daichi	Japan	10 m	L, 24 cm	4	?
Yaogan-1/3/5/8	China	5 m?	X, 3 cm	2.7	?
Risat-2	India	1 ½ m	C, 5 ½ cm	0.3	?
Lacrosse (fleet of four)*	USA	1 m?	X, 3 cm	14.0	6,000

* Exclusively military. Frequency bands – L: ~1.3 GHz; C: ~5 GHz; X: 8–10 GHz. The weight is per satellite. Cost is of the fleet of satellites (if more than one) in millions of US dollars.

Each of the non-US systems in Table 5 will now be briefly described.

Germany's Ministry of Defense has deployed a fleet of five SAR-Lupe imaging radar satellites at the relatively low cost of $400 million. In its spotlight mode of imaging, the ¾-ton SAR-Lupe provides 50-cm resolution. The fleet of five ensures that one of them will be close to the scene of interest – the maximum waiting time is 36 h.[47]

The German military is also a good customer of the TerraSAR-X commercial imaging radar satellite, for which 80% of the $230 million funding came from the German research ministry and the rest from the private company EADS Astrium. A second satellite called Tandem-X was launched in June 2010 with a view to flying in formation with TerraSAR-X, thus enabling precise 3D maps to be created.[48]

The Italian military (30%) joined forces with the civilian research ministry (70%) to underwrite the $1.4 billion needed to finance the fleet of four COSMO/SkyMed imaging radar satellites. At almost 2 tons each, these satellites provide imagery with resolution approaching that of Germany's SAR-Lupe. The fourth satellite is due to be launched in 2010.[49]

The first of Canada's Radarsat series of imaging radar satellites was launched in

---

[47]   de Selding (2004).
[48]   de Selding (2007b).
[49]   Lardier (2007), and Nativi (2007).

1994. Monitoring the ice across Canada's northern and eastern coastline is an important function of these satellites. As witnessed by the recent US contract to buy Radarsat images mentioned above, the images are obviously of interest to the military, despite the resolution being only 3 m.

One surprising omission from Table 5 is Russia. There has not been a Russian imaging radar in orbit since 1992. Two systems are under development: the Kondor-E due for launch in 2010 and the Arcon-2 in 2012; both are said to offer images with 1-m resolution.[50]

Radar images are implicitly black-and-white, but they can become "color" images of a kind. First, the radio frequency or wavelength of the radar signals varies from satellite to satellite. The Radarsat satellites use radio signals that are lower in frequency, namely "redder" that those of the TerraSAR-X satellite. The difference is roughly equivalent to that between red and green light. Thus, images of the same scene taken by these two satellites will differ in the same way as pictures of a scene in green and red light will differ – the broad outlines will be the same but some features will be brighter in one color than in another.

Another form of "color" in radar images is provided by a type of "glint" called polarization, already mentioned in Chapter 5. Polaroid glasses deal with this feature when they reduce glare, of bright sunlight reflected off water, for example – images of the same scene with and without the glasses will differ quite considerably. The radio signals of an imaging radar can be adjusted to either of two polarizations, which will change the details visible in the scene. It's not quite the same thing as color but it influences the brightness and contrast across the image.

Choosing a radar with the right frequency and then selecting the appropriate combination of polarizations can help the analyst spot ships at sea or tanks in a forest or recently disturbed earth or even to see beneath the surface (if it's dry, as in a desert). At Europe's military space imaging center near Madrid, radar imagery is used to support allied forces in Afghanistan and other trouble spots. The manager there responsible for these support activities, Jean-Charles Poletti, says that the polarization of the radar images, the width of the scene they capture and the time between images (the "revisit time") are more important than resolution.[51]

The great advantage of imaging radar is that it sees at night and through clouds. However, objects in a radar image are subtly different from an optical or infrared image and it takes experience to get full value from them.

For example, a smooth surface such as calm water appears black because the radio signals are reflected away from the satellite – the surface could be bright yellow but still show on the radar image as black. Smooth, dry sand usually gives the same effect.

Hills appear bright on the side facing the satellite and dim on the side that faces away. The side facing away from the satellite could be in bright sunlight but it will

---

[50]   Lardier (2010c), and Grishin (2009) p. 15.
[51]   *Space News* (2010a).

still appear dim. The subtlety of radar imaging is that you can see shadows such as the dim hillside just mentioned; however, the shadows are not caused by sunlight, but are caused by the radio signal transmitted by the satellite – just like you have a shadow from a streetlight on the side away from the light.

Buildings, bridges and other man-made objects contain flat surfaces that reflect radio signals strongly and therefore appear bright. Cars, trains, ships and other metallic vehicles show up well in radar images.

Features near the size of the wavelength of the radio signals, including ripples and waves on water, reflect strongly and appear bright. We saw in Chapter 5 that the longer wavelength, 23 cm, of Japan's scientific Daichi imaging radar satellite was preferable for detecting subtle subsidence of the ground at a mud volcano in Indonesia. Table 5 shows the wavelength used by each satellite – all use one of three wavelengths: 3, 5½ or 23 cm.

A mix of optical and radar images might offer the best of both worlds during the daytime – the recognizable characteristics of the ordinary optical image with features sensitive to radar highlighted. The company providing Germany's TerraSAR-X radar images to the US military offers this blend of images, since the same company operates the French SPOT satellites.

MISSILE EARLY WARNING

If you are concerned that an enemy will launch missiles at you, then with a suitably placed satellite, it should be fairly easy to spot the launch because of the brightness of the missile's rocket engine.

The satellites used to spot missile launches are called early warning satellites and the first US ones were given the deliberately uninformative title of the Defense Support Program (DSP) satellites. They were designed to watch out for an enemy launching a missile at the USA in anger and to assist in verification of arms control treaties. More than 20 countries now have ballistic missiles in addition to the USA and they perform about 100 missile test flights a year. The US response to this threat includes satellites to detect and locate missile launches, allowing radars to then target them precisely and anti-missile missiles to intercept them (hopefully). The Missile Defense Agency has been formed to lead this activity (Figure 110).[52]

The first DSP was launched in November 1970, although a series of experimental satellites, with the designations MIDAS and RTS, had been launched throughout the 1960–1966 period to demonstrate the concept of launch detection and try out various techniques – different orbits, sensors, data recovery methods and so on. The general idea was to watch for the bright flash of light and flare of heat given off by a rocket motor. They proved very successful at spotting Soviet and Chinese missile launches – by the end of June 1973, the DSP satellites had detected 1,014 launches and they added another 982 in the ensuing 18 months. This vast quantity of data

---

[52]   Covault (2007a).

Figure 110. "Ballistic missiles pose a growing, potentially catastrophic threat": Lt General Patrick O'Reilly of the US Missile Defense Agency. Credit: Department of Defense.

provided useful information to the teams of analysts trying to piece together the missile development and deployment plans of the Soviet military. DSP data allowed the launch site to be pinpointed to within 3–15 km and the launch heading to within 5–25°, depending on various factors such as the relative location of the launch and the DSP satellite. There were a few false alarms, but only for smaller missiles such as submarine-launched missiles in the northern hemisphere summer (due to the glint of the sun on the sea).

Any bright flash triggered the DSP sensors, but software algorithms sorted out the missile launches from ammunition dump explosions, forest fires, gas pipeline fires, the burn-up of satellites re-entering the earth's atmosphere, military jet aircraft using afterburner – even the July 1996 explosion of TWA Flight 800.[53]

In addition to sensors to detect the bright trace of a rocket launch, DSP satellites carried special instruments to spot nuclear explosions, taking over the role initially performed by the US VELA satellites dedicated to nuclear explosion detection. The GPS navigation satellites (see Chapter 6) are now the carriers of these nuclear explosion detection instruments. The specialized instruments on VELA, DSP and

---

[53]   Richelson (1999) pp. 65, 69, 109.

now GPS detect X-rays, gamma-rays, neutrons and the flash of light given off by an above-ground nuclear explosion, and do so very reliably. The one exception was the event recorded by an ageing VELA satellite in the early hours of September 22nd 1979 about 2,500 km off the coast of South Africa. The detectors on the two DSP satellites that had a view of that region failed to spot the event, aerial sampling missions by the Air Force and laboratory analysis of foliage from southern Africa found no evidence, while signals picked up by US Navy acoustic listening posts and ionospheric disturbance measurements by the Arecibo radio telescope were inconclusive. The USA declared the event to have been a false alarm, but recent re-evaluation of the evidence has tended to favor the conclusion that it *was* a nuclear explosion.

A nuclear explosion starts off with a short very intense flash, then a short dimming while expanding gas overtakes the fireball and masks it from view until the gas has thinned and cooled enough for the fireball to be visible again. This unusual dual flash was indeed seen by the Vela satellite, but was slightly abnormal, which was enough to convince the official US Government panel that it might not have been from a nuclear explosion. The advocates of the explosion scenario claim that the South Africans towed an Israeli nuclear bomb out to sea on a raft, waited until there was a very heavy rain storm and set it off. The storm washed the radioactive fall-out into the sea, thus explaining why the US aerial sampling missions failed to find traces of it and why the DSP satellites failed to detect it. There was certainly collaboration on nuclear bomb-making between Israel and South Africa, but whether that link-up included testing a nuclear weapon is still unclear.[54]

China is modernizing its missiles so that each carries multiple warheads, extending the range of its submarine-launched missiles and developing cruise missiles. Of the 100 or so missile tests detected by the US DSP satellites each year, about half are said to be Chinese.[55] China has deployed two new long-range missiles recently. One, the DF-31 or CSS-10 Mod 2, is fired from a fixed launch site and its range of more than 11,000 km puts most of the continental USA and Europe within its reach. The other, the DF-31A or CSS Mod 1, is launched from a trailer bed and thus movable and difficult to target. Its range of more than 7,500 km includes the US west coast and parts of Europe. China's sole XIA nuclear-powered submarine, capable of launching ballistic missiles, will soon be joined by two or more new-generation JIN boats. The new JL-2 missile has been successfully tested and when deployed in the new submarines, its range of over 7,500 km will bring parts of the USA within reach without the need for the JIN to travel far from the Chinese coast. Each submarine can carry 12 missiles. China also has a large and recently upgraded arsenal of short and medium-range missiles, most of them transportable and thus elusive. China's missiles currently carry only one warhead accompanied by decoys and penetration aids to confuse missile defense systems, but as already noted, it is in the process of

---

[54]   Richelson (1999) pp. 74, 79–81, 104–105, 108–109, and Reed and Stillman (2009) pp. 177–181.

[55]   Covault (2007a).

Table 6. US intelligence view of short-, medium- and long-range and submarine-based missile arsenals of emerging countries.[56]

	Missile	Launch type	Range (km)	No. of launchers
China	DF-11/15	Road-mobile	300–900	> 200
	DF-21/A	Road-mobile	1,800 +	< 130
	DF-3	Transportable	3,000 +	5–10
	DF-4	Silo/transportable	5,500 +	10–15
	DF-5A	Silo	13,000 +	˜20
	DF-31	Road-mobile	7,000 +	< 15
	DF-31A	Road-mobile	11,000 +	< 15
	JL-1	Submarine	1,600 +	12 not yet deployed
	JL-2	Submarine	7,000 +	12 not yet deployed
Egypt	Scud	Road-mobile	300 +	< 25
FSU[57]	Scud	Road-mobile	300	< 100 each country
India	Agni-I	Road-mobile	700	< 25
	Agni-II	Rail-mobile	2,000 +	< 10
	Agni-III	Rail-mobile	3,000 +	Not yet deployed
	Sagarika	Submarine	300 +	Not yet deployed
Iran	Various	Road-mobile	150–500	< 100
	Shahab-3	Road-mobile	1,300/1,900	< 50
	New	Road-mobile	1,900 +	Not yet deployed
Libya	Scud	Road-mobile	300 +	< 100
N Korea	Scud B/C	Road-mobile	300–1,000	< 100
	No Dong	Road-mobile	1,300	< 50
	New	Mobile	3,200 +	In development
	Taepo Dong 2	Unknown	5,500 +	Not yet deployed
Pakistan	Ghaznavi/Shaheen-I	Road-mobile	400 +	< 50
	Ghauri	Road-mobile	1,300	< 50
	Shaheen-II	Road-mobile	2,000 +	Unknown
Saudi Arabia	DF-3	Transportable	2,800	< 50
Syria	Scud D	Road-mobile	700	< 100
Vietnam	Scud	Road-mobile	300	< 25
Yemen	Scud	Road-mobile	300 +	< 25

upgrading some of them so that each carries multiple warheads able to hit several widely spaced targets – up to 1,000 km apart.

Besides China, other emerging nations deploying or testing ballistic missiles include India, Pakistan, Iran and North Korea, as shown in Table 6.[58]

Since the first was launched in 1970, 23 of the 2½-ton DSP satellites (Figure 111) have been placed into geostationary orbit that gives them continuous viewing of the

---

[56] NASIC (2009).
[57] FSU = Former Soviet Union countries: Belarus, Kazakhstan, Turkmenistan, Ukraine.
[58] NASIC (2009), and Norris and Kristensen (2008).

Figure 111. The earth-pointing telescope of the DSP satellite stands out clearly in this artist's impression, with the electricity-generating solar arrays at the rear. Credit: Northrop Grumman.

areas of interest – the 23rd, launched in late 2007, has apparently broken down and is drifting around the geostationary arc, requiring commercial satellite operators to keep an eye on it and be ready to take evasive action if it drifts too close to one of theirs. The USA made an unprecedented effort to diagnose the fault in this DSP by using a pair of miniature satellites called Mitex (that were launched a year earlier) to rendezvous with the DSP in its geostationary orbit. So far, attempts to fix it have failed, but the ability of the USA to sneak up close to a satellite 36,000 km out in space has been noted by its enemies with some trepidation – the Mitex satellites are too small to be detected by radars or telescopes on the ground.[59]

The limitations of the DSP were shown up starkly in the first Iraq conflict in 1991. Scud missiles launched by Iraq against Saudi Arabia and Israel were detected by DSP, but no launchers or personnel were ever captured or destroyed. The fault

---

[59]   Brinton (2008), and *New Scientist* (2009b).

was not the DSP's alone, but also the fact that it took perhaps 20 min for allied aircraft to reach the detected launch site, by which time the mobile Scud launcher was long gone. The weather didn't help by being the worst for over a decade, with much more cloud cover than normal for that time of year.[60] The DSP takes a snapshot of the earth below every 10 s. If it's a clear day, DSP might at best get 8–10 sightings of the rocket plume before the rocket burns out and the missile begins its coast phase – its path is said to then be "ballistic" rather than "powered" because it is influenced only by earth's gravity and it is much more difficult to spot, since its rocket motor has stopped. If there is cloud below, the DSP may only get three or four sightings once the missile has broken through the clouds before the rocket motor burns out. Each sighting is not an image, but a location of the center of the flash. Whether clear or cloudy, calculating back from the few sightings to the launch location is not easy.[61]

The intention had been to create a next generation of early warning satellites comprising a constellation of about 20 satellites in a low orbit plus four satellites in high orbits – two of the four in geostationary orbit and the other two in elliptical orbits. The high satellites spot the bright rocket flame as the missile accelerates, then the low orbit satellites are intended to track the missiles after their rocket motor has burnt out and they enter the ballistic unpowered phase of their journey. Although the satellites in low orbit would have been relatively small (half a ton or so) and thus not too expensive, the sheer numbers required made the concept a costly one and it has been put on hold. Two demonstrators were in fact launched in 2009, aimed at testing technology for a possible future operational system.

The elliptical orbits mentioned above give coverage of launches from northern polar regions. The need for this was emphasized as recently as 2009, when two Russian Sineva long-range missiles were test-fired from a submarine near the North Pole. The first missile was aimed at a Russian test range 4,000 km away in the Kamchatka peninsula on the Pacific, while the second followed a low-flying trajectory (presumably simulating how to evade enemy radars) ending up 2,500 km away in the Kanin peninsula on the other side of Russia roughly to the north of Moscow.

Russia maintains a variety of land-based long-range missiles – over 1,000 nuclear warheads on 300 missiles of six types (the USA has 450 missiles, each with one to three warheads). In addition, Russia's 10 active nuclear-powered ballistic missile submarines (the USA has 14) carry 160 missiles bearing 576 warheads between them. Three of a new generation of submarine are being tested or constructed but the new missile they will carry failed its flight tests in 2008 and 2009. When deployed, the submarines will each carry 16 missiles, each of which can carry six warheads. Despite this powerful force and continuing investment, Russia's nuclear arsenal is significantly reduced compared to its Cold War strength (see Figure 112). Russia has built up a force of 76 bombers that can carry nuclear-armed so-called cruise

---

[60]   Richelson (1999) pp. 168–173.
[61]   Richelson (1999) p. 65.

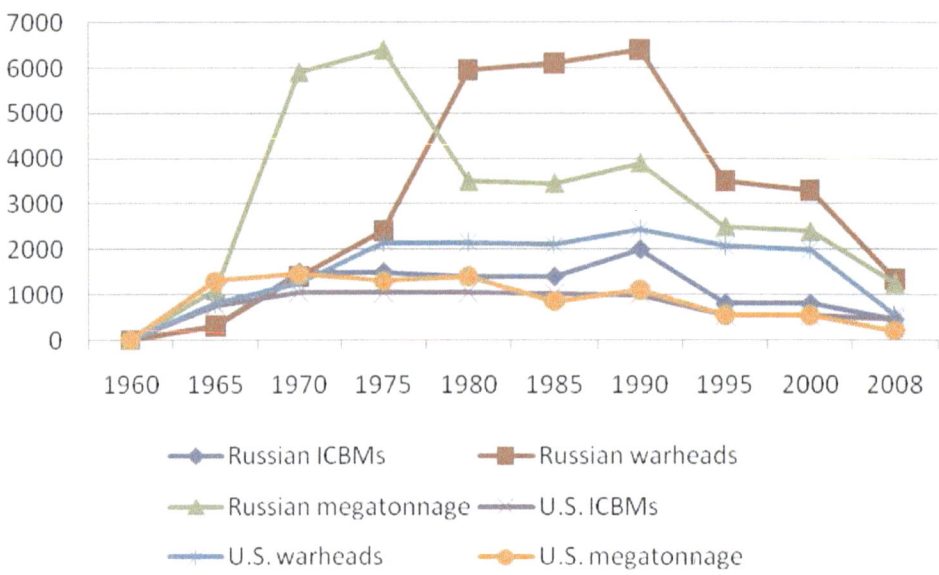

Figure 112. Trends in deployed US and Soviet/Russian ICBMs (long-range missiles) – this diagram excludes submarine-based and airborne nuclear weapons. The USA and Russia have reduced deployed nuclear weapons by about 75% since the end of the Cold War.[62]

missiles (a fancy name for unmanned aircraft), while the USA has 94 long-range bombers that make up the third leg of the US nuclear triad.[63]

In April 2010, Presidents Obama and Medvedev agreed to cut their nuclear forces further. The detailed agreement is complicated, involving a somewhat torturous mix of deployed and not-deployed weapons, warheads versus missiles, which aircraft are counted (since virtually any aircraft could theoretically carry a bomb) and so on. A limit of 1,550 deployed warheads (bombs) was agreed, which is about a third below that in the previous Bush–Putin agreement of 2002. In a separate statement, President Obama promised that the USA would no longer install multiple warheads on long-range missiles – but the wording limits this concession to land-based missiles, so submarine-launched missiles will still be able to carry them.

Negotiations continue between the USA and Russia to lower still further the number of nuclear warheads and missiles in their arsenals.[64]

---

[62] Diagram is after Norris and Kristensen (2010); Gates (2010) p. 5.
[63] The triad comprises ICBMs, SLBMs and the bombers; 18 of the bombers are the B-2 stealth bomber; the other 76 are B52-H bombers; Gates (2010) pp. 22–24.
[64] Norris and Kristensen (2010), and Bombeau (2010).

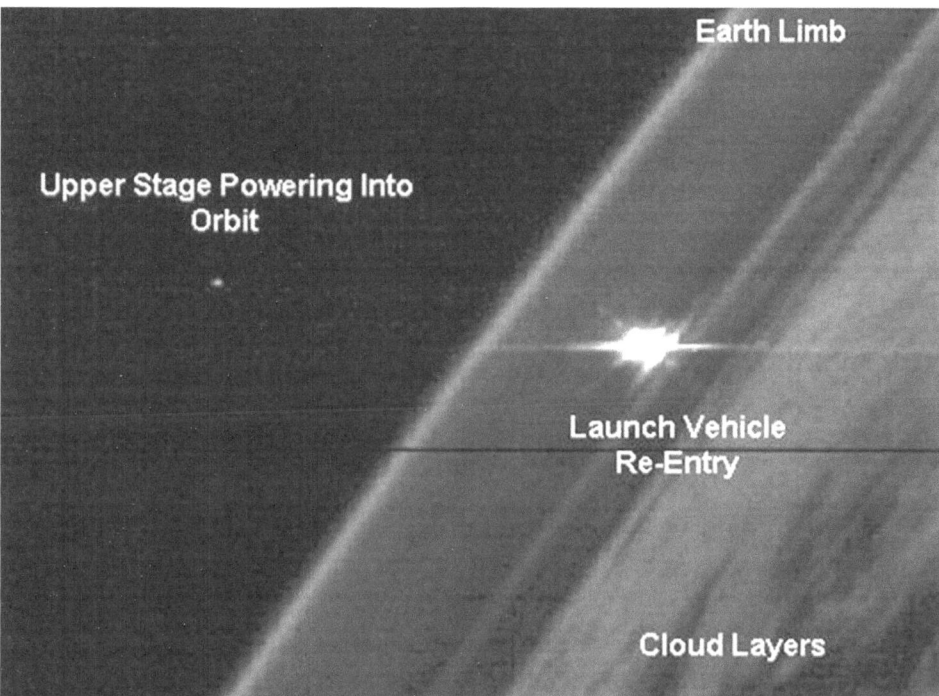

Figure 113. Image taken by the first SBIRS sensor in its high elliptical orbit showing a satellite launch. Credit: US Air Force.

The replacement for the DSP missile warning satellites has now begun to take shape. The first two versions of the highly elliptical part of what is now called the Space-Based Infra-Red System (SBIRS) constellation were launched in 2006 and 2008, piggy-backing on other unidentified (for security reasons) US military satellites, and may well be the source of the information on the Sineva launches from the Arctic mentioned above.

The technology used in SBIRS is much more powerful than in DSP. The sensor sweeps across the earth below every second or so and if it spots an event of interest, another sensor dwells on that event to pick up as much detail as possible. Unlike DSP, the output of the sensor is an image, not just a location, so further analysis can be performed on it. Figure 113 shows an image from the first SBIRS sensor launched in 2006. SBIRS was tracking a scheduled satellite launch as part of its test campaign and the small dot in the upper left shows the upper stage of that launcher. The bright flash mid-right is the lower stage of the launcher, which has been jettisoned and is re-entering the atmosphere. As with other SBIRS images released by the Air Force, stars have been blacked out so that the location of SBIRS in orbit can't be deduced. The image has also been degraded to hide the true resolution of the sensor. The sophistication of the facilities on the ground has also been enhanced – more powerful computer processing to analyze the information and faster relaying of results to the troops, ships or planes that can do something about it. Even before the first SBIRS

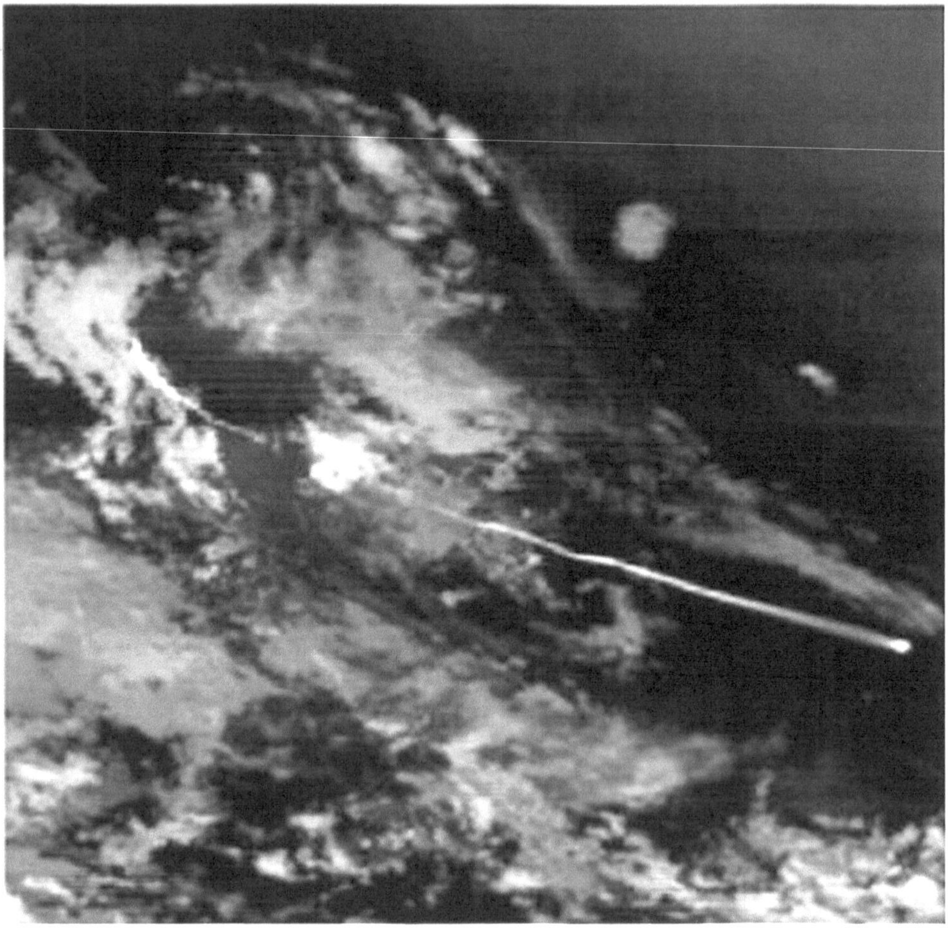

Figure 114. Infrared satellite imagery from the second SBIRS sensor in an elliptical orbit depicting a missile launch through the clouds. Credit: US Air Force.

was launched, the timeliness and quality of DSP information had improved because of this. One industry official claims SBIRS will be 10 times better than DSP, able to detect dimmer and more fleeting targets, and presumably the effects of the upgraded ground facilities will improve that even further.

The second SBIRS piggy-backing on an unnamed military satellite is also sending back high-quality information. Figure 114 shows the track of a missile above the clouds as seen by SBIRS, indicating that the trajectory can be spotted and analyzed in quite some detail.

Already 7 years behind schedule, the first geostationary SBIRS satellite fell foul of a very modern problem in 2009 in that its software was found to be faulty, resulting in a delay and modification that cost $750 million. The faults were so bad that USAF Colonel Robert Teague said "the [software] design and architecture had fundamental flaws. The solution essentially required starting from scratch". The

Figure 115. Gen. Robert Kehler is "not optimistic or pessimistic" about the trouble-plagued SBIRS. Credit: US Air Force.

first of two is now due for launch in 2011 but may start with some software still not fully tested. General Robert Kehler, the head of the USAF Space Command (Figure 115), sounded resigned to even more delays and cost increases when in the fall of 2009, he said that "I don't know if there will be any other impacts to schedule and cost. I'm not optimistic or pessimistic. We are where we are with SBIRS". What is now considered to be the full constellation of two geostationary and two elliptical satellites costing about $15 billion is due to be operational in 2014.

A back-up plan is underway in case the geostationary SBIRS hits more problems. In 2011, about the same time as SBIRS is due for launch, a commercial telecommunications satellite will carry into orbit the prototype of a special US military camera that is a potential alternative to the SBIRS technology.[65]

Russia has a mix of geostationary and elliptical orbit missile detection satellites. Since 2006, three Oko satellites have been placed in elliptical orbit, with a fourth and final one due to be launched in 2010, and one has been placed in geostationary orbit. More than 100 of these early warning satellites have been orbited since the first in 1972, initially only in the elliptical "Molniya" orbit (see below) and followed in 1975 by a geostationary orbit version. In geostationary orbit, these 2½-ton satellites can see north and south to the edge of the Arctic and Antarctic Circles. They scan about 60% of the area visible below every 7 s (versus the 10 s of the USA's DSP).

---

[65] Butler (2006), Cáceres (2006), Richelson (1999) pp. 225–226, Lardier (2006), Brinton (2009a), Brinton (2009c), and Butler (2010a).

A new system called EGS is under development and is scheduled to be launched in 2012.[66]

GETTING THE IMAGES BACK QUICKLY

As mentioned in Chapter 1, a surveillance satellite is a bit like a cell phone with a camera. You take a photo with your cell phone and then send it to someone who will be interested in it – your cell phone uses radio signals to do this, of course. Similarly, a surveillance satellite takes an image of a scene below it and then has to find a way to send it to the operator on the ground. Chucking the camera or the film out of the satellite and retrieving it on the ground is one way that is still occasionally used by Russia and China. But the obvious way is to send the image to the ground via radio.

Have you ever wanted to send a message or call someone from your cell phone only to find that you are out of network coverage? You keep the image in the camera's memory and wait until you get a signal again – and perhaps complain later to the cell phone operator in the hope that he will extend the coverage before the next time you are in that place. Surveillance satellites come across this problem quite a lot, and by and large, the solution is to store the image in memory and wait until the orbit brings the satellite over a suitable station. The USA and Russia have built special relay satellites very high in the sky so that the surveillance satellite can see one of them no matter where it is. The relay satellites are either in a geostationary orbit that is always 36,000 km above the equator or an orbit that takes it further north at a similarly high altitude. Japan, France and Germany have also experimented with relay satellites but have not yet committed to operational versions. Unless you need the images very soon after they have been taken, relay satellites are an expensive luxury – you need only wait a few hours before the satellite comes within range of a friendly ground facility and can send its images down. In addition to the expense of the relay satellites themselves, you have to add in the equipment on the surveillance satellites to communicate with the relay satellites.

The US military and intelligence communities have two sets of relay satellites at their disposal. First is NASA's Tracking and Data Relay Satellite System (TDRSS), which was mentioned in Chapter 1. The Lacrosse imaging radar satellites are thought to use TDRSS to get their images back home quickly but other US military satellites such as the Keyhole series use a military version of TDRSS called Space Data Systems (SDS) or QUASAR. Five SDS have recently been orbited into a mix of geostationary and inclined elliptical orbits. Two, in 2000 and 2001, were placed in geostationary orbit from where, in principle, they can see satellites as far as 70° north and south, namely to the edge of the Arctic and Antarctic regions. However, beyond about 50° north or south, the surveillance satellite in low orbit has increasing difficulty to stay locked on the SDS satellite as the line of sight gets closer and closer

---

[66]   Lardier (2010c), and Lardier (2008e).

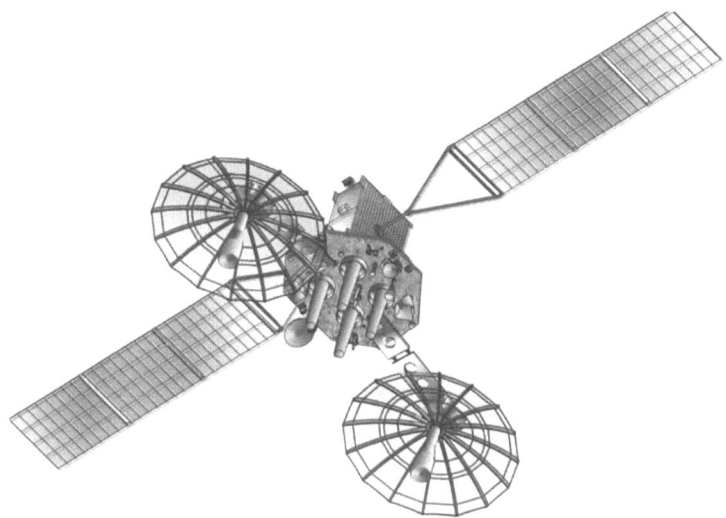

Figure 116. Russia's equivalent to TDRSS: the Loutch-5a satellite, due for launch in 2011, will provide continuous communication with the International Space Station. Credit: Roscosmos.

to the horizon – by the time the surveillance satellite is at 70° north (or south), the SDS satellite appears very close to the earth and is difficult to track. For this reason, some of the SDS satellites are placed in what is often called a Molniya orbit, named after the Soviet communications satellites that popularized the orbit 40 years ago. A Molniya orbit is high in the sky over the northern hemisphere and low over the southern hemisphere. The Soviets chose this orbit to provide broadcasting and communications to the whole of the Soviet Union, many of whose inhabitants live far to the north. The US military chose it for the same reason – to give good coverage of all of the Soviet Union, especially the militarily interesting regions around Murmansk (69° north) and the Bering Strait (67° north) next to Alaska. Out of its 12-h orbit, 8 h will be in the high northern part, so three such satellites can provide continuous 24-h coverage between them. Three SDS satellites launched in 1998, 2004 and 2007 are in these Molniya orbits.[67]

The Loutch (or Luch) series is the Russian equivalent of TDRSS (Figure 116). They provide communications with the International Space Station and with the Russian Soyuz capsule that from 2011 will be the only vehicle for transporting astronauts to and from the Station. Following a 16-year break in launches, two small (1.2-ton) versions are due to be launched in 2011 and a larger (3.2-ton) version in 2013. Once these are in place, Loutch will for the first time occupy its three designated locations on the geostationary orbit. Loutch also handles communications with other Russian satellites in low orbit – both civil and military.

---

[67]   Richelson (2008) p. 198, and Lardier (2010c).

The Potok or Geizer series are dedicated military relay satellites in geostationary orbit, working, for example, with the Yantar spy satellites and the signal eavesdropping satellites (see next chapter). The two most recent were placed in orbit in 1995 and 2000. A new generation called Garpoun is scheduled to begin replacing Potok in 2010.

China launched its first geostationary relay satellite, called Xichang, in 2008. Like US and Russian relay satellites, Xichang is intended to provide communications with astronauts – in China's case, with the Shenzou space capsule. Without Xichang, a Shenzou capsule is in communication with Chinese ground controllers only 14% of the time (when passing close to China), but with the addition of Xichang, this increases to 50%.[68] Whether China's surveillance satellites can relay data through Xichang has not been made clear.

UNMANNED AIRCRAFT

The modern way to address the need for tactical imagery is to use remotely controlled aircraft that take images of the ground below them and radio them back either directly to the forces on the ground or via satellite to a special ground terminal. The US military has taken recently to calling them remotely piloted vehicles, and that is indeed an accurate description of how they work – scaled-up versions of radio-controlled model airplanes. The US military had fewer than 50 of these machines in 2000, but almost 7,000 at the end of 2009, and Congress approved $6.1 billion to procure more of them in 2010.

These drones come in several shapes and sizes. At one extreme, they look just like a model airplane straight from ToysRus that can be operated by the local forces, taking off from a small strip of flat ground or even hand-launched, reconnoitering the terrain within a few kilometers of the take-off point, sending images back to a vehicle-mounted terminal and able to land in the same short strip of flat ground. One such is the Battlefield Air Targeting Micro Air Vehicle (BATMAV) that weighs 450 g (1 lb) and carries two cameras, one looking to the side, the other forward. Also called Wasps, these tiny spies fly short distances to provide around-the-bend and over-the-hill information. Another is the 2-kg (4-lb) hand-launched Raven, with a TV camera and an infrared camera to see in the dark and allegedly all-weather – although a strong gust of wind or a heavy rain shower would surely test its resilience. The 18-kg (40-lb) ScanEagle is launched from a small slingshot device, can travel up to 8 km and can swivel its turret to point its optical and infrared cameras as required. The US Army and Marines operate 17 different types of remotely piloted vehicle. They have names such as Gnat, Hunter, Shadow and Sky Warrior, some of them capable of working more than 100 km from base. Besides being used to see round corners, they assess damage after a raid and help target the raids in the first place.

---

[68]   Lardier (2008d).

At the other extreme is the US Global Hawk vehicle – to give you an idea of its sophistication, the over-run in the budget to develop an enhanced version (just the over-run, mind) is $2 billion – the enhanced version will carry a ton and a half of surveillance equipment instead of just a ton in the current version. This is a serious airplane. It is the size of a small commercial airliner, takes off from an airport runway and flies for up to 48 h non-stop, and over intercontinental distances if required. Global Hawk is not controlled by the man in the front line, nor by the man in the local command center. It is controlled by specially trained staff (a mix of pilots and image analysts) from Beale Air Force Base in California – even if it's flying over Afghanistan or Somalia 15,000 km or more away. In the next room is the control center for the U-2 aircraft, with human pilots that still fly reconnaissance missions in Iraq and other regions of interest – the U-2 is still in use because it can carry a more varied suite of sensors than can the Global Hawk at the moment. Images (optical, infrared and radar) are beamed to Beale from the craft via satellite or they can be transmitted directly to troops on the ground below. Commands to the craft are sent from Beale to the vehicle via communications satellite.

In between the small, locally operated unmanned aircraft and the sophisticated Global Hawk are a range of medium-sized vehicles. The Predator is one such vehicle (see Figure 121 in the next chapter) and can not only spy out the lie of the land beyond the next hill; it can do something about it if it finds an enemy there. Unlike the passive Global Hawk, which stays at high altitude (up to 20 km or 65,000 ft), Predators can get closer to the action and carry missiles that can be fired at a target in its field of view under the control of an operator. The newer Predator-B and its upgraded variant, the Reaper, can fly higher (up to 15 km, 50,000 ft) and carries more and better weapons and cameras. They fired more than 200 missiles in Afghanistan during 2009 and 69 in Pakistan.

The Predator or Reaper is controlled by operators who watch video imagery taken by it and sent back over a satellite link – more than 400 h of imagery per day in Afghanistan in early 2010. The amount of video sent back will soon be vastly increased when new versions of Reaper take to the air, each equipped with 12 video cameras that can look in all directions (the camera suite is called the Gorgon Stare) – and in 2011, the number of cameras on the latest Reapers will increase to 30! The operators control the vehicle by radioing commands over the satellite link to its flight computer. Typically, the operators are military pilots with years of experience of flying conventional aircraft. Given the thousands of remotely piloted vehicles in use today and their continued increase in numbers, it should come as no surprise that the USAF is now training more pilots to fly them than to fly traditional airplanes. The need for pilots is accentuated by the fact that these unmanned craft can fly until the fuel runs out without the need to sleep or eat. Thus, one such remotely controlled plane needs two or even three pilots while it is in the air. One counter-balancing factor is that the risk of pilots losing their lives has been eliminated.[69]

---

[69] Léon (2010), Canan (2010), Brinton (2010e), Valpolini (2010), Butler (2007a), Butler (2007b), and Butler (2007c).

In the example of the squad commander wanting to cross a ravine above, a remotely controlled aircraft is more likely to have been the imaging vehicle used were it needed than a satellite.

Unmanned aircraft beam back huge quantities of imagery, much of it in the form of video. Satellites don't usually provide video because they typically zoom across the sky from horizon to horizon in a few minutes. Those that can keep a place in view for a long period are in a very high orbit and thus unable to provide detailed imagery.

With plenty of imagery available from an unmanned aircraft, the next problem is finding a trained analyst to figure out what's going on in the picture. Off-the-record briefings by US and UK military officials indicate that only 10% of the imagery that is taken is analyzed. "We are swimming in sensors and drowning in data", quipped one. The following anecdote told to me by a senior British officer illustrates why an untrained person may make bad decisions based on video.

A group of allied forces in (say) Afghanistan is peering at a screen that is displaying video from an unmanned aircraft a few miles away. They spot a group of armed men dressed in local attire walking down a mountain path in enemy-held territory and decide to call in an air strike to attack them. The one trained image analyst among them says "Hang on! Those guys don't walk like locals". The group turned out to be an allied special forces team. The story is told to illustrate the usefulness of video, the need for trained analysts and the hopelessness of trying to automate the task of image interpretation – we are a long way from having software that can analyze cultural factors such as a man's style of walking.

The use of unmanned aircraft to follow and then kill Taliban and Al Qaeda personnel is discussed further in the next chapter.

9

Military radio surveillance from space

INTRODUCTION

When you make a phone call, the signals that leave your phone travel along copper wires or cables to the local exchange if you are using a land line or by radio link to relay towers if you are using a cell phone or by radio link to a WiFi router, cable or wire if you are using the internet. The signals then are routed by the phone company or internet service provider to their destination through cables, wires, radio links or whatever route is cheapest and least congested. If the caller and receiver are far apart, the signals may travel by microwave radio link or via undersea cables or by radio through satellites. You have no control over what route your call takes and, indeed, you have no idea what that route is – it will vary from time to time, depending on congestion across the phone network. You may get a clue that a satellite is involved if there is a slight time delay between your remarks and the response of the person on the other end. This is often seen in TV news shows in which the reporter in the field stands silent for a moment after the newsreader in the studio has posed a question. We may think the reporter is hard of hearing but it is more likely that the newsreader's question is delayed by half a second or so as it wends its way through a satellite 36,000 km above the earth.

Phone calls that involve radio links are naturally open to eavesdropping, since a person with a suitable radio receiver can pick up the signal if they are in the right area. Picking up signals that travel through wires or cables is inherently more difficult, since you have to be physically in contact with the wire or cable, or at least very close to it. The laws against listening in to radio signals vary from country to country, and in the case of the USA, as we will see below, the issue of unlawful intercepts of messages has become highly political.

P. Norris, *Watching Earth from Space,* Springer Praxis Books,
DOI 10.1007/978-1-4419-6938-5_9, © Springer Science+Business Media, LLC 2010

SATELLITES BEING LISTENED TO – AND LISTENING

The USA and a few other countries take advantage of the open nature of radio signals to listen in to those of other countries – and sometimes in their own country, too. Satellites are involved in two distinct ways in this. The first technique is to eavesdrop from the ground on conversations that are routed through commercial satellites. In the 1970s, a large proportion of international telephone calls went via satellite and this eavesdropping was very effective. In one intercepted conversation, a diplomat was heard to tell his home base that he was going to send further communications in the post, which was bad news for the eavesdropping listeners. By the 1980s, however, most international telephone traffic had switched to undersea cables. This change from satellites to cable was not an attempt to avoid the eavesdroppers, but was simply due to economics. Certain types of conversation still go by satellite, such as where one of the parties is on a ship or an airplane, and eavesdropping of these calls is still viable.

It's pretty easy to pick up calls that are routed through a satellite – the phone company has to broadcast the call from a ground antenna pointed at the satellite and receive the other half of the call the same way. Thus, the eavesdropping agency just sets up a suitable radio receiver near enough to that of the phone company to be able to listen in to the radio signals up to and down from the satellite. Listening in to an undersea cable is not so easy and requires some sort of physical contact with the cable. The relevant US agency (the National Security Agency, NSA, or the Federal Bureau of Investigation, FBI) can get a warrant to listen in to cables that come into the USA, but not for cables elsewhere in the world. The beauty of intercepting conversations that went via satellite is that there was no physical contact, and thus no way for the phone company to know if its signals were being intercepted – anywhere in the world.

The USA and its English-speaking close allies, the UK, Canada, Australia and New Zealand, have collaborated to listen in to commercial and military satellites around the world. Ground stations located in those countries pick up communications from dozens of satellites and channel them to the headquarters of the NSA in Maryland, USA. The activity became public under the name ECHELON, although, in fact, ECHELON is only the name of a software program that sorts the information.[1] This eavesdropping was heavily criticized by the European Parliament in 2001, not only because of the invasion of privacy involved, but also because of suspicion that the information was used to help American business. The European Parliament's criticism was tempered by the fact that "only a very small proportion of communications are transmitted by satellite". The Parliament noted that "the majority of communications [can] be intercepted only by tapping cables and intercepting radio signals".[2] The continued extension of undersea cables means that

[1]   Richelson (2008) p. 229.
[2]   Schmid *et al.* (2001), p. 133.

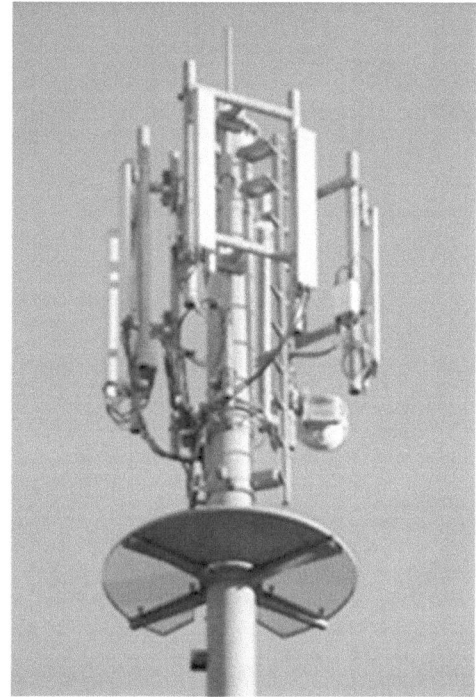

Figure 117. A microwave tower (left) that sends bulk communications over long distances and a cell phone tower (right) that relays cell phone calls to and from the rest of the network.

fewer and fewer countries rely on satellites for international communications. The countries that lack a cable connection and therefore need satellites are mostly African or isolated small islands, such as in the Pacific Ocean.

The second general way to eavesdrop from space is to have special satellites that listen to normal radio communications. In the 1960–1990 period, US agencies had satellites to listen in to two main types of radio communications: military signals (to be discussed below) and long-distance commercial communications. The satellites that undertake this sort of eavesdropping are still highly classified and not much has been published about them. The commercial communications were those between microwave towers that proliferated during the 1960s and 1970s (Figure 117). The telephone operators transmitted long-distance telephone and fax messages between the towers before routing them into the copper wires that link each home and business to the telephone network. It was a lot cheaper and faster to build microwave towers to carry this traffic than to string wires or cables over hundreds of miles. As with any radio broadcast, the telephone company's receiver on its tower was not the only device able to receive the communication – anyone else's receiver could, too. The eavesdropping satellites would pick up what is sometimes called microwave spillage – the signals intended to go between the microwave towers but in fact available for anyone with a suitable antenna to receive.

Figure 118. GCHQ's new headquarters in Cheltenham in the west of England. Costing over $600 million, it was opened by Queen Elizabeth II in 2003. Credit: Crown Copyright.

The first US communications intelligence (COMINT) satellite was launched in 1968. Called CANYON, it was the first of a series that is thought to continue today, although the name has changed over the years – to CHALET, then VORTEX, then MERCURY. The name changes were to some extent caused by security lapses such as that by British spy (and pedophile) Geoffrey Prime. Prime was a section head at the UK equivalent of NSA, the Government Communications Headquarters (GCHQ) in Cheltenham, England (Figure 118), and was privy to intercepts from CANYON and other systems and to the code-breaking abilities of the USA and UK. He was eventually stopped because his wife informed the British police about his child abuse activities, whereupon his spying work for the Soviet Union was unearthed. He went to prison in 1982 and served half of his 38-year prison sentence before being released in 2001.

The realization of the information Prime had fed to the Soviets strained NSA's relationship with GCHQ. Were it not for the fact that a US citizen, Christopher Boyce, had also been caught selling intelligence secrets to the Soviets a few years earlier (discussed further below), the USA–UK cooperation agreement might have been severed.

The technology of these eavesdropping satellites is impressive. They are located 36,000 km out in space and appear more or less stationary from the ground below. A satellite at 200-km altitude circles the earth in about 90 min. Satellites farther out take longer to do so, culminating at 36,000 km, when a satellite takes 24 h to complete its circuit of the earth, meaning that its movement is synchronized with that

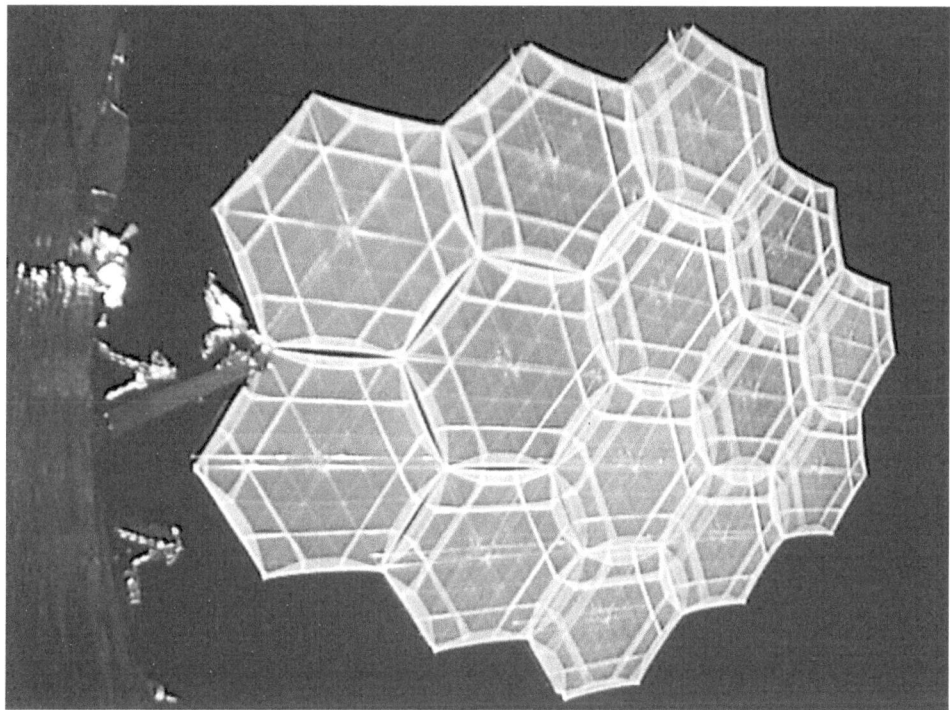

Figure 119. Photograph taken on December 26th 2006 of one of the two 19 × 17-m antennas unfurled in geostationary orbit on Japan's Kiku-8 experimental telecommunications satellite; the camera is attached to the main body of the satellite and took images of the antennas as they went through a complicated deployment sequence. Courtesy JAXA.

of the earth below.[3] Being so far from the ground, the satellites need enormous radio antennas to pick up radio signals. The largest antennas on civilian satellites are currently the two tennis-court-sized 19 × 17-m antennas on Japan's experimental Kiku-8[4] satellite launched in late 2006 (see Figure 119) and the 18-m (60-feet)-wide circular dish on the Terrastar-1 satellite launched in July 2009 (see Figure 120). But the secret eavesdropping satellites are thought to have already been that size 40 years ago, to have been double that 10 years later, 50 m wide by 1994 and 90 m by 2006[5] – at least one newspaper report claimed they had dishes 150 m wide (almost 500 ft).[6]

In the 1980s and then more rapidly in the 1990s, long-distance communications grew rapidly as email, internet and other digital communications caught on. Instead of building more microwave towers, phone companies found it made economic sense

---

[3] Satellites higher than 36,000 km take longer than 24 h to circle the earth.
[4] Previously called ETS-8.
[5] Richelson (2002) p. 157, Bamford (2001) pp. 369, 401, and Heyman (2009).
[6] Johansen (1995).

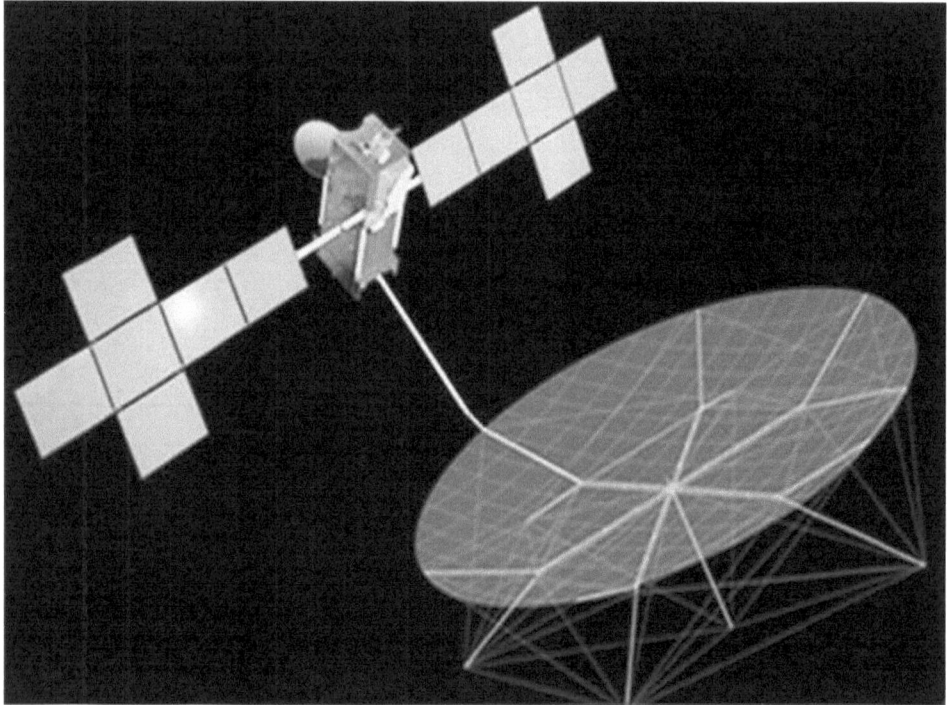

Figure 120. Artist's impression of the Terrestar-1 communications satellite with its 18-m antenna deployed; the antenna is folded during launch. The large antenna is needed because its users have small cell phone-sized terminals. Credit: Space Systems/Loral.

to install fiber-optic cables between the main cities, so the eavesdropping satellites found they had less and less to listen to.

Another trend during this period was for communications to become digital (that is to say, able to be processed in a computer) rather than analog. With digital came encryption – if you can process the information in a computer, you can easily encode it to make it difficult for an eavesdropper to read. The US eavesdropping agencies resisted this trend by forbidding the telephone or internet companies from using the most sophisticated encryption. The encryption that was permitted can be "cracked" by the agencies, although the agencies claim that it is costly to do so – you need super-fast (i.e., expensive) computers. A spate of recent stories in the technical press has suggested that there may be fast and cheap methods to read the encrypted messages and if academic mathematicians are finding that out now, perhaps the eavesdropping agencies knew it all along.[7]

With the growth of cable on land and under the sea, communications were becoming harder for the eavesdropping agencies to intercept until along came the mobile phone explosion in the late 1990s and once again, communications were

---

[7]   Kleiner (2009), O'Brien (2009), and Vijayan (2010).

being transmitted over the air. There are about five billion mobile phones in use around the world and the number keeps growing. All of the chatter, text messages, email, downlinks, etc. from mobiles goes by radio to the relay masts that have sprung up by the million across the globe. US and perhaps Russian satellites are, in principle, able to listen in to the billions of conversations and messages coming from or going to cell phones but the scale of doing so is daunting. Before the eavesdroppers can decide whether a conversation is worth listening to, they have to decrypt it, so they need some clue as to which of the billions of calls are worth devoting time and effort to. The number of the caller and the number of the receiver are not encrypted, so one tactic is for an eavesdropper to wait until a telephone number of interest is involved in a call before bothering to decrypt. Another tactic might be to focus on calls from or to a particular area.

One example of how this information is used by anti-terrorist authorities has been well publicized.[8] The NSA spotted a phone call coming from a phone number on its watch list. It belonged to Qaed Salim Sinan al-Harethi, who was believed to be the Al Qaeda operative who had planned the attack on the *USS Cole* in a Yemeni harbor in 2000 that killed 17 US sailors. The call was probably using an Inmarsat satellite phone and NSA was probably picking up all Inmarsat signals in the region at a suitably located ground facility. Al-Harethi's phone contained a GPS chip and this enabled the NSA operators to determine that the phone was in the Marib Province of Yemen, al-Harethi's native country. A mountain range that rises to 3,760 m (12,340 ft) separates Marib province from the coast and the capital, explaining why Marib is largely beyond the control of the Yemeni Government. Al-Harethi was high on the USA's terror wanted list, so the CIA operation in nearby Djibouti was alerted. An unmanned Predator aircraft similar to that in Figure 121 was quickly directed by the CIA to the Marib area, with the intention of killing al-Harethi on sight. However, the NSA analyst was puzzled. He had been listening to al-Harethi on tape for several years and could tell that he was not the speaker on the phone. Then, the speaker engaged in a conversation with someone else in his vehicle – asking directions from a man in the backseat. The analyst recognized the backseat person as al-Harethi and this was confirmed by one of his colleagues listening to a playback of the short conversation between driver and backseat passenger. With this confirmation that al-Harethi was in the vehicle, the Predator was ordered to fire a Hellfire missile, which destroyed the car and all of its occupants.

Another example is the interception of a satellite phone call being made by Nek Mohammed in June 2004 in Pakistan. Leader of the militants in the border area of South Waziristan, Nek Mohammed had fought alongside the Taliban in Afghanistan and had helped Al Qaeda forces to escape from the Tora Bora mountains into Pakistan after the USA-backed forces swept the Taliban out of power in 2001. In early 2004, his forces inflicted a humiliating defeat on a Pakastani army offensive in the region. Thus, the USA was quick to order a missile strike once a Predator had

---

[8]   Bamford (2006), and Bamford (2008) pp. 135–136.

Figure 121. Predator in flight carrying two miniature surveillance craft under its wings. It has an endurance of 40 h and a satellite data-link system, and carries two color video cameras, infrared camera and imaging radar. Credit: General Atomics.

locked onto his satellite phone.[9] A US airstrike that killed Taliban leader Mullah Akhtar Mahammad Osmani in Afghanistan in 2007 was also based on intercepting his satellite phone call.[10]

The USA is not alone in targeting enemies by listening to their satellite phone calls. Russia is said to have killed Chechen rebel leader Dzhokar Dudayev in 1996 by using such a call to pinpoint his position.[11]

MILITARY

So far in this chapter, we have addressed eavesdropping on commercial telephone and internet traffic. The military, too, have a plethora of satellites for listening in on their enemies – actual and potential.

The first satellite to "watch" the earth was in fact one such – the tiny GRAB satellite (Figure 122). GRAB was launched into orbit in 1960 and provided hitherto inaccessible information about radar systems deep inside the Soviet Union. GRAB worked by receiving radio signals across a very wide range of frequencies and relaying them to a station on the ground. The satellite didn't process the signals in

---

[9]   Rashid (2009) p. 272.
[10]   Smith (2006).
[11]   Hendrickx (2005) p. 99.

Figure 122. The world's first spy satellite: the US GRAB ELINT. Credit: National Security Agency.

any way, just acted like a "bent pipe" in taking radio signals that it heard at its 1,000-km altitude and beaming them downwards to a friendly station. By having a station suitably located near the border, signals from 5,500 km inside the Soviet Union could be picked up. Signals from two Soviet radar systems, one associated with SAM-1 anti-aircraft missiles, the other with missile early warning, were detected and could then be analyzed. The advantage of knowing the characteristics of an adversary's radar was not only that you could in future know what it was when you detected it, but you could design electronic counter-measures to jam it or confuse it. GRAB was the first of a long line of what became known as Electronic Intelligence (ELINT) satellites.

The success of GRAB encouraged the USA to use satellites extensively for analyzing its enemy's radars and other military signals. Radio transmissions from Soviet missile tests to their ground operators had been monitored from stations in friendly countries that bordered the Soviet Union, such as Turkey and Iran, and on ships and islands in the Pacific Ocean, where many of the long-range missile tests ended. As explained by British intelligence expert Chapman Pincher,[12] "the signals emanate from devices fitted to various points on the missiles to inform engineers on the ground about velocity, altitude, aerodynamic details and engine performance,

---

[12]   Pincher (1984) p. 559.

which they must know to improve range and accuracy". Such data also measure temperatures, electrical currents, fuel usage and the on/off status of each piece of equipment. The data[13] are particularly valuable if a missile fails in some way, since, otherwise, the engineers will be left guessing as to the cause of the failure. The data are just like the information in an aircraft black-box – a critical source of information for air crash investigators after an aircraft accident. The challenge for the eavesdropper is to figure out what the data mean. The task has been likened to trying to understand the switches and dials on a car dashboard if all words and numbers have been deleted from them. By a process of informed guesswork, you would gradually work out which data refer to speed and what the units are, which to engine status and so on. Satellites ensured that data from missile tests right across the Soviet Union and the Pacific Ocean could be intercepted.[14]

Communications between Soviet military units were another target for interception – they used radio channels different from commercial traffic, hence the need for dedicated listening equipment.

The tiny GRAB evolved into massive satellites with enormous antennas, as discussed above. Many of these satellites were placed in geostationary orbit, 36,000 km above the earth, hence the need for their giant antennas. One series already mentioned above, initially targeting communications intercepts, began with the launch of CANYON in 1968,[15] which evolved a decade later into CHALET, and then endured name changes due to security leaks into VORTEX and then MERCURY. A second series that was targeted at data from missile tests began in 1970 with the launch of the first RHYOLITE satellite (later renamed AQUA-CADE). This series evolved to pick up communications transmitted across the Soviet Union via microwave towers.[16] It was succeeded in 1985 by the first of two MAGNUM satellites (renamed to ORION). As we approach the present day, details about the current US eavesdropping satellites become more and more vague. A new generation of satellites began to appear in 1994, with code names such as RAMROD, RUTLEY and RANGER.

In addition to satellites in geostationary orbit, another group of eavesdropping satellites has been picking up radio signals in what is called a Molniya orbit, discussed in the previous chapter – this is an elliptical orbit that is at its maximum height over the northern hemisphere and at its lowest over the southern. Their maximum height is similar to a geostationary satellite with a correspondingly broad view of the earth below. The big advantage is that they cover northern latitudes better than from geostationary orbit but the disadvantage is that they move along their orbit from north to south and thus are over the northern area of interest only

---

[13]   Often referred to as "telemetry" (which literally means "measurement at a distance") or "telemetry data".

[14]   Richelson (2002) p. 86.

[15]   The names of US spy satellites are always written in upper-case – the reason for this pretentious habit is not known to the author.

[16]   Urban (1996) p. 57.

for eight or so hours a day. Three satellites in Molniya orbit are sufficient to provide continuous coverage over a given area. Initially called JUMPSEAT satellites, these evolved into the TRUMPET satellites that were launched in 1994, 1995 and 1997. The first in what is said to be a TRUMPET replacement series was launched in 2006.[17]

There are thought to be about a dozen US ELINT satellites currently in orbit. The present generation are sometimes called INTRUDER or PROWLER in the press, of which one, with the designation NRO-26, was launched in January 2009 and another is thought to be NRO-32, due to be launched in late 2010.[18]

MONITORING THE OCEANS

Besides air defense radars, missile data and military communications inside the Soviet Union, another target for US ELINT satellites was Soviet ships. Out in the ocean, far from the radio-rich environment on land, ships are, in principle, easy to spot – they will be the only source of radio signals for miles around. Ships transmit radio signals to communicate with each other and with their base. Their radars also transmit very strong radio signals (radar is an acronym loosely derived from "radio detection and ranging") and then detect the much weaker echoes as the radio signals bounce off objects such as other ships, airplanes, missiles or objects on shore. The beauty of this situation is that even if you can't decipher the radio signals, you know a ship is there. Both the USA and the Soviets quickly grasped the potential of this scheme and placed in orbit groups of two, three or four satellites so that they could triangulate the signals from a ship using the different time and direction of the ship's radio signals as seen by each satellite in the group. The series of US satellites had friendly names (at least for use publicly) starting with Poppy in the 1960s, then White Cloud in the 1970s and 1980s. More recently, these satellite groups are referred to as NOSS (Naval Ocean Surveillance Satellite) followed by a -1, -2 or -3 to signify which generation (White Cloud being NOSS-1). The next NOSS to be launched is said to be the satellite labeled NRO-25, due for lift-off from Vandenberg Air Force Base in California in mid-2010.[19] A pair of NOSS satellites, each weighing about 3 tons, plus the rocket on which they are launched into orbit costs about $600 million – there are thought to be 20 in orbit altogether, spaced so as to give as near as possible continuous coverage of the ocean areas of interest. A database of all ship movements is maintained based on NOSS information plus data from Navy and Coast Guard air and sea surveillance. Civilian as well as military ships are tracked, reflecting the threat from terrorists and the interest in the navies of countries such as China and Iran.[20]

---

[17]   Richelson (2008) pp. 217–219.
[18]   Lardier (2009f), and Lardier (2009d).
[19]   Lardier (2009d).
[20]   Covault (2007b), and Covault (2007c).

In addition to detecting ships and working out their exact location by triangulation, these satellites allow the ships' radar signals and radio communications to be collected for later analysis. In recent years, the US military have acknowledged that specific launches are in the NOSS category but not that they are a group – civilian observers have detected additional objects close to the acknowledged one, but the military refer to those objects as "debris". The current NOSS-3 satellites come in pairs, as has been recorded by many ground-based observers just by watching them with binoculars. The satellites are about 1,000 km high and one of the pair trails the other by about 250 km. The height is a compromise between being as close to the earth as possible in order to detect faint emissions and being as high as possible to get a broad view. The relatively small separation of the pair ensures that they both see the same target, thus allowing the signals to be compared and the target's location and direction to be triangulated.

If the enemy ships are careful and observe radio and radar silence when the NOSS satellites pass overhead, it would be useful if the satellite had a radar of its own that could detect the ships below. There are occasional press and web reports that NOSS possesses such radars but little is known about them. Other reports say that NOSS satellites can detect the faint radio emissions from a ship's engines and the magnetic effects of a ship's hull.

SPYING ON THE SPIES

The name changes of the various US ELINT programs were sometimes because the public became aware of the previously classified name. Unfortunately, the Soviets also got hold of information about them from time to time. One instance was when NSA and GCHQ agreed to collaborate on the CANYON program. In return for getting access to the data from CANYON, GCHQ provided analysts to NSA to help with the analysis of the vast amounts of data it was intercepting. One of the GCHQ analysts was Geoffrey Prime, who was identified as a Soviet spy in 1984 (see above). Prime gave details of the CANYON program to the Soviets.[21]

The RHYOLITE program suffered in the same way. Christopher Boyce was an employee of TRW, the company that manufactured the RHYOLITE satellite. He and his friend Andrew Lee sold information about RHYOLITE, its proposed successor called ARGUS and the KH-11 optical surveillance satellite (see Chapter 8) to the Soviet Union from 1975 to 1977. These guys were not criminal geniuses – Lee was arrested by Mexican police when he tossed a heavy object over the wall of the Soviet Embassy in Mexico City. The Embassy thought it was a terrorist bomb and summoned the local police, who found a strip of microfilm containing classified documents inside the object (a Spanish–American dictionary). Lee had wanted money urgently to fuel his drug business. He and Boyce were put on trial in the USA

---

[21]   Richelson (2002) p. 157.

Figure 123. Soviet spy Christopher Boyce, apprehended in 1981 after his escape from prison. Credit: Department of Justice.

and received heavy prison sentences – life for Lee, 40 years for Boyce.[22] The story of Boyce and Lee was popularized for a mass audience in Robert Lindsey's 1979 bestseller *The Falcon and the Snowman* and the 1985 film of the same name, with Timothy Hutton as Boyce and Sean Penn as Lee and directed by John Schlesinger. A sequel to the story was the 19-month period of freedom enjoyed by Boyce in 1980–1981 after he escaped from the Lompoc Correctional Institution in California. He was eventually apprehended by Federal Marshals near Seattle (see Figure 123).[23]

These breaches in ELINT security probably allowed the Soviet Union to mislead and confound US analysts for many years – false radar signals, bogus test data from missiles, changed encryption schemes and the like. The possibility that the Soviets were falsifying the test data transmitted from their missiles had major repercussions. Once it became clear in the USA that this might be happening, many US officials became skeptical about making further agreements with the Soviets to limit strategic weapons (nuclear bombs, long range missiles, etc.) – the concern was that the Soviets

---

[22] Richelson (2002) pp. 203–205.
[23] Boyce was paroled in 2002 and released in 2008. Lee was paroled in 1998.

would spoof US attempts to monitor Soviet missile tests,[24] which is an important element in the first strategic arms agreement. Matters are said to have improved after 1979 when 30 years of trying finally bore fruit and NSA cracked the Soviet military communication codes.[25]

MONEY NO OBJECT

The cost of these satellites apparently is astronomical. Press reports of the January 2009 launch of the 6-ton NRO-26 referred to it as a "highly upgraded ORION electronic eavesdropping satellite" and said that "the combined cost of the spacecraft plus booster is roughly $2 billion".[26] Recent cost over-runs of several types of US spy satellites have become public knowledge (see Chapter 8) and ELINT satellites are no exception. An attempt to move to a more cost-effective approach was discussed in the trade press for several years and would have involved a larger number of cheaper satellites – but it seems to have been a failure, since the large expensive type is still being launched every 2 or 3 years.

House of Representatives Intelligence Committee Director John Millis is quoted as saying that "We spend more money on one satellite in one year than we do on all the analytic capabilities combined". It's difficult to turn this remark into a cash figure, since the statements about overall NSA budget range from $7 billion in 2000–2001 to near $40 billion in 2007.[27] There is ample room for officials to be disingenuous or confused about NSA budgets, since the costly satellites can be considered as part of the Department of Defense or the National Reconnaissance Office budget when that's convenient. However you interpret these figures, since there are at least 2 years between the launch of each of these satellites, a price tag well in excess of $1 billion is plausible.

The high cost of these monster satellites means that they are designed to last for 15 or more years – it would be too expensive to replace them more frequently. But communications is an area that changes dramatically every few years, so the expensive long-lasting satellites risk becoming obsolete before they have served their term. One dramatic change that is perplexing the eavesdroppers is the emergence of telephony via the internet – Voice over Internet Protocol (VoIP), to give it its full name. Lost within the cacophony of data flowing through the internet, VoIP phone calls are hard to detect. The internet breaks up the data into "packets" and sends them to their destination via any of thousands of routers (a fancy name for computers that route the data). A packet for a phone call is indistinguishable from an email packet or a web page packet. Worst of all, some packets in a conversation might go via one router, other packets via other routers – they get put back together

---

[24]   Pincher (1984) p. 563.
[25]   Bamford (2001) p. 370.
[26]   Covault (2008b).
[27]   Bamford (2001) pp. 466, 482, and Bamford (2008) p. 199.

Figure 124. NSA Director Lt General Keith Alexander exploits the US position as the home of the internet. Credit: National Security Agency.

again only when they reach the recipient of the call. So, eavesdropping on part of the internet is only likely to give you at best a part of a conversation. Sir David Pepper, who was GCHQ's Director from 2003 to 2008, called the move to VoIP as "his biggest problem. It is a complete revolution; the biggest change in telecoms technology since the invention of the telephone".[28]

The good news (for the USA) is that a large proportion of the world's internet traffic is routed through the USA, even if sender and recipient are elsewhere in the world – that's the benefit of inventing the system and being its largest supplier of equipment. NSA Director Lieutenant General Keith Alexander (Figure 124) admitted that one of NSA's "greatest advantages [is] the ability to access a vast portion of the world's communications infrastructure located in our own nation".[29] A subtle but apparently persistent problem is that NSA is forbidden from listening to US citizens without prior authorization. The problem is that to tell whether one of the speakers is American, you have to listen to the conversation – in his 2008 best seller, *The Shadow Factory*, James Bamford has described how this Catch-22 tied NSA in knots for several years, as they veered between intercepting almost nothing (thus being legally watertight) to intercepting nearly everything in the post-9/11 environment (arguably illegal).

---

[28]   Bamford (2008) p. 221.
[29]   Bamford (2008) p. 207.

THE SOVIETS/RUSSIANS: OCEAN SURVEILLANCE

The Soviet Union needed to detect American ships more than the USA needed to detect theirs throughout the Cold War. The America Navy had ruled the waves since World War II, while the Soviet Union was largely surrounded by land and frozen water, and thus naval power was never a high priority for them. At its peak during World War II, the US Navy had over 100 aircraft carriers – to get some idea of the significance of this figure, consider that in Britain and France today, the decision to build a single aircraft carrier is a political decision taken very rarely and at Prime-Ministerial level. The US Navy in 1944 was larger than the combined strength of all the other navies in the world. This titanic naval armada was the reaction to the December 7th 1941 Japanese attack on the US fleet in Pearl Harbor. The Head of US Naval Operations was the "profane, intemperate, womanizing, Olympian autocrat" Admiral Ernest King, who was described by his daughter as "entirely even-tempered; he was always angry". He took full advantage of the unlimited funds available to build ships and, according to one respected historian, created a navy "whose size owed little to rational assessment of the resources needed to defeat Japan, and almost everything to his own grandiose vision".[30]

Although the US Navy was cut back after World War II, it still remained a formidable force and a cause for serious concern to the Soviet Union. In the 18th and 19th centuries, the expansion of the Russian Empire was often described as an attempt by Russia to gain warm water ports. In its original form, Russia's coastline gave access to the sea only in the narrow confines of the Baltic Sea near St Petersburg and in the Arctic. By the 20th century, Russia had reached the Pacific Ocean, but only in the relatively unfriendly waters of eastern Siberia, where the largest port was Vladivostok. The popular quiz question to name the most southerly of Venice, Vienna, Vancouver and Vladivostok relies on the tendency to equate southerly latitude with warmth. However, in this instance, the relationship breaks down, so that Venice, at 45° latitude, has a balmy Mediterranean climate while the sea in Vladivostok harbor, 2° further south, regularly freezes over in winter[31] and thus is not a reliable base for a nation's strategic naval forces.

As the Cold War developed from the late 1940s onwards, the USA gradually built up alliances with countries along the Soviet border – for example, West Germany and Greece in Europe, Turkey and Iran in West Asia, Taiwan, South Korea and Japan in East Asia. US bases and listening posts in those countries, including 80 radar stations in Europe alone, plus the vast US fleet fed Soviet paranoia about being surrounded – and just because they were paranoid didn't mean they were wrong.

Rather than try to match the USA at sea, the Soviets focused on anti-ship missiles in order to keep US fleets at bay. They were carried by a mix of submarines, destroyers and patrol boats, and would fire them at an enemy over the horizon and

---

[30]   Hastings (2008) pp. 21, 101–103.
[31]   Vancouver is at 49° latitude and Vienna (Austria) at 48°.

out of range of the Soviet ship's own radars.[32] As part of this policy, the Soviets somehow needed to be able to locate US ships. Initially, this was performed using long-range radar-carrying aircraft, but these were limited in their range and at risk of being shot down.

As the space age gathered pace in the 1960s, the Soviets were quick to create a series of satellites to monitor US naval movements. Starting in 1967, they launched satellites that carried radars to detect US ships even when their radios and radars were switched off. Radars involve *transmitting* a radio signal and listening for echoes, and transmitting calls for a lot more power than does listening – compare the tens or hundreds of kilowatts your local TV station uses to transmit its signal versus the watts or milliwatts your TV or 3G cell phone needs to receive them.

The Soviets decided to produce electrical power for their satellite using a small nuclear reactor. A nuclear reactor has the advantage of working in the absence of sunlight – a situation that low-orbiting satellites experience for as much as a third of the time. Most satellites get around the absence of sunlight by carrying large batteries that are charged up by solar cells when the sun is visible and discharged in darkness. The radar needed too much electrical power for solar cells to be a practical option in the 1960s and 1970s, so a reactor fuelled by 31 kg of uranium 235 (^{235}U) was used to generate about 5 kW of power. The satellite as a whole weighted nearly 4 tons, of which about 1 ¼ tons was the reactor assembly.[33] The disadvantage of this approach is that if the satellite crashes, you create radioactive debris. As an aside, note that this awkward fact had persuaded Western nations to avoid the technology, except for space probes heading far out in the solar system, where there is no other option because sunlight is too weak – and even these missions provoke public demonstrations because of the risk of a crash, if the launcher explodes, for example. As mentioned before, radar satellites need lots of electrical power, so the Soviets decided to take the risk of a crash. They increased the risk by flying the RORSATs in an orbit about 250 km high. This relatively low orbit gave better results with the radar (the closer you are to the target, the better) but meant that the satellite would be dragged down within a few months by the thin extension of the atmosphere at that altitude, eventually crashing to earth. They tried to minimize the chances of the satellite crashing unexpectedly by boosting it from its low orbit up to a so-called "disposal orbit" above 900 km after its operational life of a few months was completed. At that altitude, it was calculated that it would remain in orbit for at least 400 years, by which time the radioactive fuel would have decayed naturally and be relatively harmless. A further safety measure was to have the reactor eject all its fuel so that the uranium would not be protected by the thick reactor and would burn up if it re-entered the atmosphere.

The general public became aware of Soviet radar satellites when Cosmos 954 crashed in Canada's North West Territories in January 1978, spreading radioactive debris over a wide area (Figures 125 and 126). The Soviets had to compensate

---

[32]   Siddiqi (1999) p. 398.
[33]   Siddiqi (1999) p. 402.

Figure 125. Two Canadian officials hold radiation detectors while the third digs for Cosmos 954 debris. Credit: Natural Resources Canada.

Canada to the tune of several million dollars for the cost of the Cosmos 954 clean-up, but this didn't stop them using the same radioactive technology on later satellites. There were two other unintended RORSAT crashes – a launch failure in 1973 sent the reactor to the bottom of the Pacific Ocean north of Japan, and a re-entry due to a failed attempt to reach its disposal orbit in 1983 saw the fuel ejected automatically and burn up, while the reactor and the satellite ended up in the Indian Ocean.[34]

Note that Cosmos 954 and similar satellites, which were called Radar Ocean

---

[34]   Siddiqi (1999) p. 412.

Figure 126. Piece of radioactive Cosmos 954 debris found in northern Canada. Credit: Natural Resources Canada.

Reconnaissance Satellites (RORSATs) in the West, used radar in the traditional sense, not for imaging. Their main objective was to pick up the clear radar echoes produced by shipping. The blip of an aircraft on an old-fashioned circular air traffic control radar screen gives an idea of the information these satellites provided. This form of radar requires much simpler electronics than the imaging kind mentioned in Chapters 4 and 8.

The RORSAT series is called "active" because the radar transmits radio signals to bounce off targets below. A series of "passive" satellites that listen without transmitting has also been used by Russia to detect ships for over 30 years, picking up ELINT. These ELINT Ocean Reconnaissance Satellites (EORSATs) worked in harmony with the RORSAT vehicles, which is why they were never merged into the Tselina series described below.

The last RORSAT was launched in 1988. Russia continues to launch ocean surveillance satellites of the passive or EORSAT type, such as the 3-ton Cosmos 2421 satellite in June 2006, whose mission ended in March 2008. This orbited at an altitude of about 410 km, whereas its US equivalents orbit at about twice that altitude (thus getting a wider view). A new generation to replace the EORSAT series is expected to be launched soon. Called Pion, an artist's impression in Figure 127 shows that it will include a larger and bulkier antenna (what looks like a wire grid stuck to the side of the satellite in the figure). Whether it will continue the tradition of putting much of the electronics in a sealed capsule that maintains sea-level air pressure is not known – use of a pressurized capsule avoids the electronics having to be specially developed to withstand the vacuum of outer space, but adds a lot to the weight.

THE SOVIETS/RUSSIANS: EAVESDROPPING

The Soviets appear to have been extremely successful at eavesdropping on US military communications during the Cold War. This was due to a mix of intelligence-gathering satellites and spies inside the US military and intelligence communities.

Like the USA, the initial Soviet eavesdropping activities were to probe electromagnetic emissions from defense systems such as radars and to listen in on communications between military forces. The Soviets do *not* appear to have emulated America's move to giant "listening" satellites in geostationary orbit (36,000-km altitude) – they have commercial satellites at that altitude to broadcast TV and the like, but seemingly not eavesdropping ones. They also have Potok-Geïzer relay satellites in geostationary orbit so that the eavesdropping satellites lower down can transmit the information they collect back to Russia, even when on the other side of the world.[35] The actual eavesdropping satellites themselves are confined to relatively low orbits less than 1,000 km high – typically about 850 km.

The current Russian eavesdropping satellite series is the Tselina-2, of which recent launches have included Cosmos 2406 in 2004 and Cosmos 2428 in 2007. A model of the Tselina-2 is shown in Figure 128 – the actual satellite weighs over 3 tons and is about 3 m (10 ft) high. The illustration shows that the Tselina-2 lacks the giant dish that dominates the RHYOLITE, CANYON and other American geostationary eavesdropping satellites. The head of the Ukrainian factory that manufactured the first-generation Tselina satellites described the two main objectives of monitoring the radars of the USA and its allies as, first, "finding out their capabilities in a combat situation and obtaining data to take [spoofing or blocking] counter-measures" and, second, "to detect signs of changes in the activities and battle-preparedness of foreign armed forces". The lack of powerful computers to process the data collected seems, however, to have limited the ability of the Soviets to achieve these objectives. The data collected by Tselina-2 are sent back to earth via the Geizer data relay satellite (see Chapter 8), which is much faster in getting information to intelligence analysts than on the previous generation's tape recorder, whose contents were radioed to earth once or twice a day.[36]

Both the Tselina-2 and the ocean surveillance EORSAT satellites mentioned above are built in the Ukraine. That was not a problem when both Russia and the Ukraine were part of the Soviet Union. Today, this and other post-Soviet space and military issues are a source of tension between the two countries. The best publicized of these issues was the expiration in 2017 of the Russian lease of a naval base at

---

[35] The Garpoun series is the new generation to replace the Potok-Geïzer series, with the first launch due in 2010.

[36] Hendrickx (2005) pp. 101–102.

Figure 127. Future Pion ocean surveillance satellite. Credit: TsNIRTI.

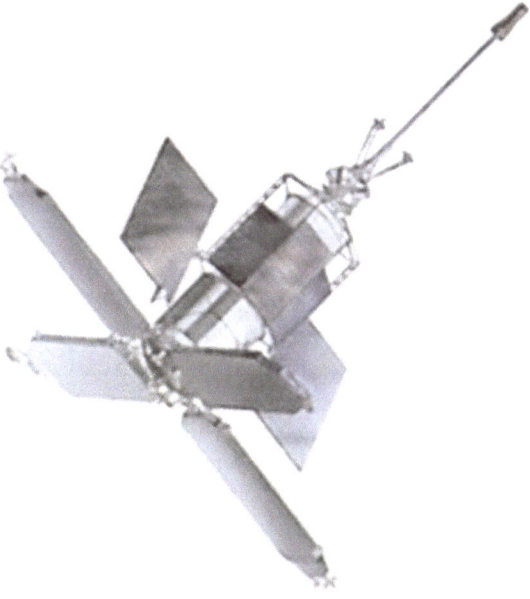

Figure 128. Tselina-2 satellite. Credit: Yuzhnoye.

Sevastopol on the Black Sea in Ukraine's Crimean region.[37] Sevastopol is Russia's only warm-water naval base and a critical Russian military asset, even though passage through Turkey's Bosphorus Strait is required to reach beyond the Black Sea to the Mediterranean and the Atlantic beyond that. As noted above, lack of access to warm-water ports has been a strategic issue for Russia for centuries. Saber rattling by both parties has been endemic since relations were strained by arguments over natural gas; Russia has occasionally briefly halted supplies when Ukraine withheld payment of increased prices. There are other awkward interdependencies resulting from the provision by Ukrainian factories of many elements of Russian rocket launchers and missiles, all potential sources of tension between the two former Soviet partners.

Following a hotly contested election victory in February 2010, the new pro-Russian Ukrainian President Viktor Yanukovych quickly negotiated a deal with Russia extending the Sevastopol lease for 25 years and getting a 30% reduction in gas prices in return. Smoke bombs and eggs were hurled and punches were traded in Ukraine's Parliament as the deal was approved by a narrow majority.

Under a November 2008 agreement, Ukraine will deliver the last four Tselina-2 eavesdropping satellites to Russia by the end of 2011. Despite the Ukrainian promise to continue supplying these spy satellites, Russia has been developing an indigenous replacement for both the Tselina-2 and the EORSAT. The first demonstration model of the Lotos that will replace the Tselina-2 was launched in November 2009. The Russian-built Pion replacement for EORSAT (illustrated in Figure 127) is still under development. Following the agreement to extend the Sevastopol lease, Russian Premier Vladimir Putin promised cooperation on aircraft manufacturing, shipbuilding and the generation of nuclear power. Spy satellites were not explicitly mentioned – in public, at least. Collaboration between Russia and the Ukraine on a civilian surveillance satellite called Sich-2 may also get a boost from the resolution of the Sevastopol issue.[38]

During the Cold War, the Soviet Union had the benefit of extensive information about US radar and military communications codes thanks to the work of their spies. Right after World War II, the USA could decipher many of the most secret Soviet communications thanks to information and equipment gained from captured German intelligence staff. Having the UN headquarters in first San Francisco and then New York made it easy for the US intelligence agencies to eavesdrop on Soviet and other countries' communications. The Soviets, on their side, were content with its location in the USA since it gave their spies an excuse to be there in large numbers.

But the story changed overnight in 1948 when the Soviets changed all their cryptographic equipment and codes, making it impossible for the USA to decipher them. The Soviets had placed spies inside at least one of the US intelligence agencies and had discovered the vulnerability of their communications. The spy or spies in

---

[37]   See, e.g., Schwirtz (2008).
[38]   Lardier (2009b), Lardier (2010c), Levy (2010), and Stern (2010).

question were never caught, although William Wiesband was strongly suspected.[39] The North Korean invasion of its southern neighbor in 1950 came as a surprise to the intelligence agencies, which presumably would not have been the case if the USA had been reading Soviet diplomatic and military communications during the preceding year – North Korea was strongly assisted by both China and Russia.

Two NSA mathematicians, William Martin and Wernon Mitchell, defected to the Soviet Union in 1960, strengthening the Soviet ability to confound US eavesdropping. The intelligence agencies were not totally in the blind in the lead-up to the Cuban missile crisis in 1962 because they could at least detect the new Soviet air defense radars installed in Cuba and could monitor the often unencrypted conversations in both Russian and Spanish of Soviet airmen as they were transferred to Cuba.

This blackout on the US ability to read Soviet ciphers continued for 30 years, until 1979. The breakthrough is said to have come as a result of small, steady and persistent advances in understanding how the Soviet encryption machines worked. Once the full breakthrough was achieved, NSA was able to "hear secure encrypted voice communications better than they were hearing each other".[40] However, before this occurred, America's most costly (in cash and lives) anti-communist conflict, the Vietnam War, had been fought and lost without the ability to intercept Soviet communications.

The situation was actually worse than this. The Soviets were relatively secure from US intercepts, at least of their most important traffic. But, in the other direction, the Soviets knew a lot about how the US systems worked. For 17 years, from 1967 to 1984, the Soviets were receiving detailed cryptographic code information from a US Navy communications officer, John Walker (Figure 129). The KGB boss in the Soviet Embassy in Washington, DC, in 1967, Boris Solomatin said in 1995 "Walker enabled your enemies to read your most sensitive military secrets; [for] seventeen years we were able to read your cables". A few months after Walker began his career as a Soviet spy, America's *Pueblo* spy ship was captured largely intact by the North Koreans, allowing the cipher machines it contained to fall into Soviet hands – the crew had failed to destroy them before being captured. Once the USA knew that *Pueblo* had been lost, all crypto keys were changed throughout the US Navy. What the USA didn't know was that one of the people receiving the changed lists was John Walker. Thus, for the next 17 years, the Soviets had the three types of cipher machine aboard the *Pueblo* plus regular updates of the codes and cryptographic keys in use.

Walker's KGB case officer Oleg Kalugin said that the twin Soviet intelligence coups (the *Pueblo* and Walker) enabled the Soviets to keep track of US nuclear submarines and naval ships. He described it as "the greatest achievement of Soviet intelligence [during] the Cold War".[41]

---

[39] Bamford (2001) pp. 23–24.
[40] Bamford (2001) pp. 359, 370.
[41] Bamford (2001) pp. 276–277.

Figure 129. John Walker – arguably the most important Soviet spy during the Cold War. Credit: Department of Justice.

And that's not all. Spies such as Geoffrey Prime in Britain and Christopher Boyce in California (see above) provided the Soviets with details of the eavesdropping satellites. It seems likely that the Soviets then took measures to ensure that the information collected by those satellites was compromised. This could include transmitting deliberately false information to mislead the US listeners, as well as the simpler action of changing their procedures and equipment to prevent the interception of signals and communications.

Data from Soviet eavesdropping satellites were down-linked to facilities in the Soviet Union and its allies, including Cuba. The Cuban facility was also a listening post in its own right – indeed, the main source of information on radars and communications on the continental USA. It was based at Lourdes just to the southwest of Havana, and thus only 150 km from Key West. It picks up communications via commercial satellites as well as microwave spillage mentioned above and signals emitted by the hundreds of air defense missile batteries across the USA. Smaller variants of Lourdes were present in Soviet embassies around the world.[42] Vladimir Putin, then Russian President, agreed to close the Cuban listening station in 2001. However, a Chinese listening station is said to be still there.[43]

As described earlier, US eavesdropping satellites in the RHYOLITE and MAGNUM series are designed to listen in to the data sent back during missile tests. Sensors on the missile that is being tested measure speed, acceleration, temperatures, electrical current, fuel, the on/off position of switches, etc. and this is radioed back to the ground so that the results of the test can be analyzed in detail.

---

[42]   Hendrickx (2005) p. 98.
[43]   Richelson (2008) p. 396.

The Soviets/Russians could collect this kind of information using ships, since US long-distance missile tests head out over the ocean – and are announced in advance to warn civilian shipping. Hence, the Soviets never built satellites dedicated to this task.

OTHER COUNTRIES

France[44]

France has been experimenting with satellites for monitoring radio transmissions for over 15 years. The activities have been funded jointly by the military and the civilian space agency and are intended to lead to an operational system called Ceres in 2016.

The first French test satellite was launched in 1995. Called Cerise (the French for "cherry"), it was a small (50-kg, 110-lbs) and low-cost satellite purchased from Surrey Satellite in the UK.[45] Cerise had the distinction about a year after launch of being struck by a piece of debris in space – later determined to have been part of a rocket launcher. The debris hit a long boom that extended from Cerise for stability purposes – the boom kept Cerise pointing towards the earth in much the same way as a trapeze artist uses a long pole to help stay upright on a high wire. After the impact, Cerise was knocked into a different pointing direction until the engineers on the ground were able to adjust its stabilization logic. It then continued working for several more years.

Cerise was followed in 1999 by another test satellite with a fruity name: Clementine. Again weighing only 50 kg, Surrey Satellite supplied only half of the satellite this time, with a French manufacturer supplying the militarily sensitive receiving equipment.

In 2004, France launched a group of four small satellites, each weighing 120 kg. Called Essaim (French for "swarm") (Figure 130) and still considered experimental, this group was a step towards an operational surveillance system in that one of the four was a back-up in case one failed, and each satellite monitored a slightly different set of radio frequencies.

The purpose of all of these satellites has been to figure out the location of and, where possible, identify individual transmitters and radars. This information would help French forces to avoid detection in any military conflict but, initially, the objective has been to understand how much information can be obtained in order to justify the funding of a future operational system. They have all been placed into orbits that are about 700 km high – thus not requiring the gigantic receiver dishes of the American eavesdropping satellites that hover 36,000 km above the earth. It is not

---

[44]   Lardier (2009a), Lardier (2008a), Taverna and Wall (2009), *Air & Cosmos* (2007a), and de Selding (2007a).

[45]   For more about Surrey Satellite, see Chapter 5.

Figure 130. The four Essaim ("swarm") ELINT satellites being inspected before launch. © EADS Astrium/D. Marques.

clear whether the goal is also to intercept individual communications or whether it is limited to identifying and characterizing the different types of transmitter.

The next step is another four-satellite group called Elisa (an acronym for ELectronic Intelligence SAtellite), due to be launched in late 2010. Costing about $150 million, Elisa is explicitly aimed at creating a map of radar and other transmitters around the world and defining their characteristics. An important goal is to provide continuity from the earlier satellites, thus detecting at least the major changes in the radio map as time goes by. There is some irony in the fact that Russia will provide the launcher for Elisa, given that 20 years ago, Russia would have been its main target. The new partnerships of post-Cold War politics are illustrated by the Russian rocket in question, a Soyuz, being launched from a French launch site: Kourou in French Guiana (South America).

France's longer-term aim is to find other European countries to help fund the operational version of all this – already given the name Ceres. Ceres was the Roman goddess of agriculture and motherly love, but it's not clear what relevance this has to the program; Ceres is also the name of the largest object in the asteroid belt between Mars and Jupiter, which is perhaps a more appropriate reason for naming a satellite after it. Advertised to cost about $500 million, Ceres has so far attracted interest from Greece and Sweden, although they have not yet made a formal commitment to it.

China

Although China joined the USA and Russia in 2003 in being able to launch humans into space, its spy satellites are considerably less sophisticated and mature than those of the two space super-powers. Its Shijian 11-1 satellite, launched in November 2009, is thought to have been an electronic eavesdropping satellite following on from four of the earlier satellites in the Shijian series with similar characteristics – three of the four were actually pairs of satellites that were launched together. Officially, all of these satellites are described as undertaking scientific and technical research.[46] My speculation that one or both of the two most recent Yaogan satellites is an electronic surveillance satellite was discussed in Chapter 8.

UK

The UK has had close ties to America in the intelligence world since World War II. As mentioned earlier in this chapter, America's NSA and Britain's GCHQ have a close working relationship that has given GCHQ an understanding of what can and can't be learned from eavesdropping satellites. Until the 1980s, the UK relied on the USA for satellites and made its contribution in the form of ground stations in Britain, Hong Kong, Cyprus, Diego Garcia (an island in the Indian Ocean) and perhaps elsewhere. The 1982 war between Britain and Argentina for control of the Falkland Islands in the South Atlantic showed Britain's complete reliance on America for certain types of intelligence. Although the USA provided significant help to Britain in the Falklands conflict, it wasn't at war with Argentina, so there were limits to the extent of the help that Britain could expect. The British contribution to the USA–UK partnership was compromised somewhat when GCHQ staff went briefly on strike and began a "work to rule" campaign,[47] which was only put to rest when British Prime Minister Margaret Thatcher made trade union membership illegal for GCHQ staff in 1984.[48]

Conscious of its dependence on American goodwill for continued satellite-based intelligence, Britain decided to strengthen its position by constructing a fully fledged RHYOLITE-type eavesdropping satellite that would make the USA–UK dependence a bit more of a two-way affair. Thus, in 1983, the UK began developing a hugely expensive satellite called Zircon. Intended for geostationary orbit 36,000 miles high, Zircon would have complemented the American satellites collecting information from Soviet and other communist transmissions. GCHQ Director Brian Tovey saw Zircon as keeping the (UK–USA) relationship sweet and getting his

---

[46] Lardier (2009e).
[47] In "work to rule" staff stick pedantically to the letter of their work orders, e.g., not re-tuning their receivers when the target signal drifted slightly.
[48] This ban on trade union membership was lifted in 1997.

organization into space.[49] A special windowless factory was built to produce the satellite and a cover story prepared claiming that it was an extra satellite in the Skynet series of military communications satellites.

Two factors doomed Zircon. The first was the increases in its price tag – from the initial estimate of $150 million, it soared to $700 million and more. Cost escalation of military systems is very common, so this fact alone might not have halted the program. The second and fatal blow was the discovery of the ultra-secret program by an investigative journalist called Duncan Campbell. The secrecy of Zircon meant that it was not reported in the usual way to Parliament. Campbell made the absence of Parliamentary knowledge of Zircon the main focus of his investigation, since the need for transparency in government was a theme of much of his writing. Of course, he found it difficult to get information on such a secret activity. He wasn't really sure that it existed until the day he phoned a senior manager in the company that was building Zircon and bluntly asked for information about it. "You're not supposed to know about that," spluttered the surprised manager, thereby confirming that it did indeed exist.[50] Once the story was published and the escalating costs became public knowledge, Zircon was cancelled.

The sequel to Zircon is that the UK then decided to help fund NSA's MAGNUM/ORION satellites in return for guaranteed access to its data. The cost of this "guarantee" is reported to have been about $750 million, so UK tax payers funded American industry to build ORION instead of British industry building Zircon.[51] Now *there's* a subject for a British investigative journalist to get his teeth into!

---

[49] Urban (1996) pp. 60–61.
[50] Private communication.
[51] Urban (1996) p. 63.

10

The future

Satellites are extensively used by military forces as one of their routine sources of information. The same is not yet true of civilian security agencies such as border guards, coast guards, immigration agencies or regular police. However, we may expect increasing use by such agencies of satellite images now that the price of the images is falling and the level of detail they contain is improving.

Some people say that we wouldn't know about climate change were it not for satellites. Satellites certainly document changes in climate and related topics on a global scale. However, the key measurement for climate change was the carbon dioxide record in the ice dug out in Greenland or measured in weather stations on top of mountains in Hawaii. The amount of carbon dioxide in the air has been growing steadily since the start of the industrial revolution 250 years ago, and it shows no signs of letting up. The rising world temperature triggered by that carbon dioxide has many uncomfortable repercussions, such as melting glaciers, rising sea levels and changes in rainfall patterns. Satellites are central to our ability to monitor these problems worldwide. The ingenuity of space engineers now allows us to watch at night and through clouds.

Satellites are good at watching other important changes such as the spread of deserts or the shrinking of lakes – they capture the big picture that can then be documented in detail on the ground. They are also objective, as when monitoring the area of land being farmed in a totalitarian state that otherwise tends to invent the statistics to suit the propaganda line.

These climate-related and man-made changes will continue to be important for the foreseeable future. Indeed, as the world's population inexorably grows (6 billion, rising towards 9 billion by the year 2100), these problems will surely intensify. Satellites will therefore be even more important in keeping us informed about the state of the world.

Increasing population means increasing pressure on the world's resources. This is clearly true for non-renewable resources such as oil and gas, minerals, topsoil and fossil aquifers. It is also true for normally replaceable resources such as fresh water and fish as they are exploited beyond their ability to recover. Satellites can often monitor these resources, but countries have to want to address the situation for it to improve.

So, the future will certainly include more and different types of surveillance satellites as the world watches us destroy the environment we live in. Or perhaps the sight of the shrinking forests and expanding deserts will trigger an era of constructive collaboration between countries to husband the resources we have.

P. Norris, *Watching Earth from Space*, Springer Praxis Books,
DOI 10.1007/978-1-4419-6938-5_10, © Springer Science+Business Media, LLC 2010

There will be more of the fleets of smallish satellites (weighing a ton or less) that provide frequent updates on what's happening. The information needed to predict climate change will be collected more systematically than now.

One cloud on the horizon is the amount of debris in space, especially in the region below 1,000-km altitude favored by most surveillance satellites. Dead satellites and launchers are now sufficiently numerous that they will inevitably collide with each other or with live satellites (as has already happened – see Chapter 8), resulting in an increase in debris even if we never launched another satellite. Action needs to be taken to remove the largest of these dead objects by sending "space tugs" to propel them into the ocean – removing four or five a year would be enough to gradually bring the debris population under control. The alternative is for future satellites to incorporate expensive protective features such as multilayer surfaces and to organize an effective debris detection and alert service.

One development that may take a while to emerge is the increased use of unmanned aircraft especially to monitor security situations. This trend is less certain because the airlines and helicopter operators don't like the idea of thousands of radio-controlled airplanes wandering through the air-lanes. A way will have to be found to bring these remotely controlled planes under the same air traffic control scheme as piloted planes. Once that is achieved, the way will be open for them to be used to monitor everything from pollution to road traffic, from icebergs to crowds. They are already proliferating among the leading military powers and that trend will continue at least in countries in which machines are cheaper than people.

Unmanned aircraft will increase the importance of satellites. They are both types of machine that complement each other – satellites giving the global picture, unmanned aircraft providing local detail. In some cases, satellites are called on to provide the local detail, such as where a country forbids any form of monitoring by land, sea or air (North Korea springs to mind as a current example).

As explained in Chapter 6, GPS and other navigation satellites are *not* watching us – the technology works by our satnav device watching them. However, the public perception that GPS "knows" where we are is likely to continue as satnav is used on more and more occasions. The deployment of fleets of GPS look-alikes by Russia, Europe and China will strengthen the take-up of the technology. It is in high-end cars now, but will be in all cars soon. It is in top-end cell phones like BlackBerry and iPhone now, but will be in all cell phones and other mobile devices soon. Today's youth seems sanguine about their views and whereabouts being known, such as on YouTube and Facebook, so the invasive nature of shared satnav information is unlikely to slow the trend down.

Google Earth and its competitors are moving towards imagery that is more up-to-date. It would not be surprising if they were to offer a service whereby you could order an up-to-the-minute image of the area of interest – for a fee. The satellite imagery would presumably be whatever was most recent and might therefore not actually be completely current. It could also link in to street-level CCTV cameras or perhaps to anyone in the area with a camera-equipped cell phone who would receive part of the fee. The technical opportunities for intrusive surveillance are far from being fully exploited.

References

Air & Cosmos (2007a) ELISA sur ARIANE-5, September 14th, **2091**, 17.

Air & Cosmos (2007b) Kobalt-M en orbite, June 15th, 152.

Air & Cosmos (2010) Kobalt-M, April 23rd, **2215**, 35.

Albright D. and Brannon P. (2010) Suspect Reactor Construction Site in Eastern Syria: The Site of the September 6 Israeli Raid? The Institute for Science and International Security, October 23rd, available online at *www.isis-online.org/ publications/SuspectSite_24October2007.pdf*.

Anthony K. W. (2009) Methane: A Menace Surfaces, *Scientific American*, December, **301**(6), 44–51.

Arnold D. C. (2008) *Spying from Space: Constructing America's Satellite Command and Control Systems* (College Station, TX: Texas A&M University Press).

Avenhaus R., Kyriakopoulos N., Michel M., and Stein G. (eds) (2006) *Verifying Treaty Compliance: Limiting Weapons of Mass Destruction and Monitoring Kyoto Protocol Provisions* (Berlin-Heidelberg: Springer).

Bamford J. (2001) *Body of Secrets: How America's NSA and Britain's GCHQ Eavesdrop on the World* (London: Century).

Bathke C. G. *et al.* (2009) The Attractiveness of Materials in Advanced Nuclear Fuel Cycles for Various Proliferation and Theft Scenarios, *Proceedings of Global 2009*, Paris, France, September 6–11th, Paper 9543.

Bamford J. (2006) "He's in the Back Seat", *Atlantic Monthly*, April.

Bamford J. (2008) *The Shadow Factory: The Ultra-Secret NSA from 9/11 to the Eavesdropping on America* (New York: Anchor Books).

Bawa J. (2010) Seeing the Wood and the Trees, in *Environment, Sustainability & the Food Challenge*, insert in *London Sunday Times*, April 18th, available online at *www.climatechangeandthefoodsupplychallenge.com/ClimateChangePages/clima-te.pdf*, 52–53.

BBC (2009) Street View under Fire in Japan (May 14th), Greece Puts Brakes on Street View (May 12th), All Clear for Google Street View (April 23rd), BBC News Online, *www.bbc.co.uk/news*.

Bell R. E. (2009) The Unquiet Ice, *Scientific American*, February, **298**(2), 52–59.

Ben-David A. (2010) Ballistic Boost: Iran Displays New Rocket Mock-Up, but

P. Norris, *Watching Earth from Space,* Springer Praxis Books,
DOI 10.1007/978-1-4419-6938-5, © Springer Science+Business Media, LLC 2010

Views Differ on its Potential Threat, *Aviation Week & Space Technology*, February 15th, 33–34.

Berfield S. (2010) Bill Bratton, Globocop, *Bloomberg Businessweek*, April 12th, 48–53.

Bethune J., Famiglietti J. S., de Linage C., Reager J. T., Lo M., Swenson S. C., and Rodell M. (2009) Water Storage Change in California's Sacramento and San Joaquin River Basins Since 2003, Including Central Valley Groundwater Depletion, *American Geophysical Union Fall Meeting*, December 14–18th, H11D-0838 (a PowerPoint version is available online at *www.nasa.gov/pdf/411644main_GRACE-AGU%20Press-Conf-2009.pdf*).

Bindoff N. L. *et al.* (2007) Observations: Oceanic Climate Change and Sea Level, in Solomon S. *et al.* (eds), *Climate Change 2007: The Physical Science Basis. Contribution of Working Group I to the Fourth Assessment Report of the Intergovernmental Panel on Climate Change* (Cambridge, UK and New York: Cambridge University Press).

Blakely R. (2008) Google Earth Accused of Aiding Terrorists, Times Online, December 9th.

Bombeau B. (2009) Helios 2B: applications opérationnelles, *Air & Cosmos*, December 18th, 54–55.

Bombeau B. (2010) Nucléaire: un bon accord, en trompe-l'il, *Air & Cosmos*, April 16th, 34–35.

Brinton T. (2008) DSP Constellation Health Concerns Prompt Plan for Gap-Filler Satellite, *Space News*, November 24th, 1–4.

Brinton T. (2009a) House and Senate at Odds Over SBIRS Follow-On System, *Space News*, October 5th, 5.

Brinton T. (2009b) Proposed Imaging Satellite Effort Draws Congressional Opposition, *Space News*, April 6th, 1.

Brinton T. (2009c) Software Fix Adds $750M to SBIRS Price Tag, *Space News*, July 27th, A3.

Brinton T. (2009d) US Loosens Restrictions on Commercial Radar Satellites, *Space News*, October 12th, 4.

Brinton T. (2010a) NGA Awards Three Contracts for Radar Satellite Data, *Space News*, January 4th, 5.

Brinton T. (2010b) NOAA Budget Request Includes Funds to End NPOESS Contract, *Space News*, March 22nd, 5.

Brinton T. (2010c) NRO Chief Aims to Restore Technology Development Funding, *Space News*, April 19th, 7.

Brinton T. (2010d) Pentagon Eyes New Imaging Capabilities, Ground Systems, *Space News*, January 18th, 12.

Brinton T. (2010e) Pentagon Needs Long-Term Planning to Support UAV Operations, GAO Says, *Space News*, April 5th.

Brinton T. (2010f) Space Hardware Firm Doubles Down on Ship Tracking Service, *Space News*, April 19th, 14.

Brinton T. (2010g) White House Dissolves NPOESS Partnership in Blow to Northrop, *Space News*, February 8th, 5.

Broad W. J. (2010) Research Reactors a Safety Challenge, *New York Times*, April 13th, New York Edition, D3.

Broad W. J. and Sanger D. (2007) With Eye on Iran, Rivals Also Want Nuclear Power, *New York Times*, April 15th, Late Edition—Final, **1**, 1.

Brown L. R. (2009) Could Food Shortages Bring Down Civilization? *Scientific American*, May, **300**(5), 38–45.

Butler A. (2006) Fresh Eyes: First Images from New Sbirs Sensor Bring Hope for the Program after Major Setbacks, *Aviation Week & Space Technology*, November 20th, 22–23.

Butler A. (2007a) Caught on Tape: Insurgents Shift Tactics in Iraq to Try to Evade Predator Sensors, *Aviation Week & Space Technology*, March 12th, 62.

Butler A. (2007b) Constantly Watching: The Global Hawk Is Coming of Age in Iraq, Afghanistan, *Aviation Week & Space Technology*, March 12th, 56–59.

Butler A. (2007c) Going Global: Remote Piloting Allows USAF to Consider Centralized High-Altitude Recon Ops from California, *Aviation Week & Space Technology*, March 12th, 60–61.

Butler A. (2010a) Missile Warning Alert: Stratcom Chief Ups the Pressure to Find Backup Plan for Sbirs Missile Warning System, *Aviation Week & Space Technology*, April 19th, 35–36.

Butler A. (2010b) Time for a Change: With GPS IIF Set to Launch Next Month, USAF Officials Embrace a Major Departure for GPS III Work, *Aviation Week & Space Technology*, April 12th, 44–47.

Butler A. and Taverna M. A. (2010) New Image: After Years of Failed Space Radar Efforts, US Turns to Foreign Commercial Providers for SAR Data, *Aviation Week & Space Technology*, April 12th, 50–51.

Cáceres M. (2006) Cost Overruns Plague Military Space Programs, *Aerospace America*, January, 18–23.

Canan J. (2009) Cloudy Forecast for NPOESS, *Aerospace America*, October, 38–43.

Canan J. (2010) ISR in Today's War: A Closer Look, *Aerospace America*, March, 30–36.

Carlson B. (2009) Keynote Speech at GEOINT Symposium 2009, San Antonio, Texas, October 21st, available online at *www.nro.gov/speeches/GEOINT_-Speech09.pdf*.

Carlson B. (2010) Keynote Speech at National Space Symposium, Colorado Springs, Colorado, April 14th, available online at *www.nro.gov/speeches/DNRO_National_Space_Symposium.pdf*.

Carroll R. (2010) Peruvian Glacier Split Triggers Deadly Tsunami, *London Guardian*, April 13th, available online at *www.guardian.co.uk/world/2010/apr/13/peru-glacier-ice-lake-tsunami*.

CGMS Secretariat (2010) *Report of the 37th Meeting of the Coordination Group for Meteorological Satellites* (Darmstadt, Germany: EUMETSAT), February 2nd, available online at *www.eumetsat.int/groups/sir/documents/document/pdf_cgms_rep37.pdf*.

Chaisson E. J. (1998) *The Hubble Wars* (Cambridge, MA: Harvard University Press).

Chen J. L., Wilson C. R., and Tapley B. D. (2009) Satellite Gravity Measurements

Confirm Accelerated Melting of Greenland Ice Sheet, *Science*, September, **313**(5795), 1958–1960.

Chesters D. (2009) GEO-NEWS AROUND THE WORLD, online compilation at *http://goes.gsfc.nasa.gov/text/geonews.html*, December 14th.

Clark S. (2010) Quiet Sun Puts Europe on Ice, *New Scientist*, **206**(2756), April 17th, 6–7.

Clissold P. (ed.) (2008) *Candidate Earth Explorer Core Missions: Reports for Assessment: CoReH₂O—Cold Regions Hydrology High-Resolution Observatory* (Noordwijk, The Netherlands: European Space Agency), ESA SP-1313/3, November.

Collins G. P. (2007) Kim's Big Fizzle: The Physics behind a Nuclear Dud, *Scientific American*, **296**(1), 8–9.

Cooper T. (2009) Pen Hadow: Snow Patrol, *The Daily Telegraph*, available online at *www.telegraph.co.uk/sport/4176711/Pen-Hadow-Snow-Patrol.html*, January 9th.

Coordinates (2010) India and Russia Plan for JV to Produce GPS Receivers, *Coordinates*, March.

Covault C. (2006) Fade to Black: US Secret Satellites Make 16 Runs a Day over Iranian Nuclear Sites, but Such Comprehensive Intel May Not Last the Decade, as Space Recon Dwindles, *Aviation Week & Space Technology*, May 15th, 24–26.

Covault C. (2007a) Eyes on China and Iran, *Aviation Week & Space Technology*, April 9th, 48–54.

Covault C. (2007b) Ocean Recons Readied: NRO Prepares Sea Surveillance Flight, Optical Satellite Procurement, *Aviation Week & Space Technology*, April 30th, 24–25.

Covault C. (2007c) Secret Maneuvers: Wayward NRO Ocean Surveillance Spacecraft Is Shifted into Slightly Higher Orbits, *Aviation Week & Space Technology*, July 23rd, 38–39.

Covault C. (2007d) Space Control: Chinese Anti-Satellite Weapon Test Will Intensify Funding and Global Policy Debate on the Military Uses of Space, *Aviation Week & Space Technology*, January 22nd, 24–25.

Covault C. (2007e) Spy vs. Spy: Israeli Space-Based Radar Set for Indian Launch as Pentagon's New Commercial Recon Poised for Vandenberg Flight, *Aviation Week & Space Technology*, September 17th, 28–30.

Covault C. (2008a) New Russian Recon, *Aviation Week & Space Technology*, August 4th, 32.

Covault C. (2008b) Night Fright: NRO Braces for Flight of Eavesdropping Spacecraft on Agency's First Delta IV Heavy, *Aviation Week & Space Technology*, December 8th, 26.

Cruz R. V. *et al.* (2007) Asia, in Parry M. L. *et al.* (eds), *Climate Change 2007: Impacts, Adaptation and Vulnerability. Contribution of Working Group II to the Fourth Assessment Report of the Intergovernmental Panel on Climate Change* (Cambridge, UK: Cambridge University Press), 469–506.

Dalton S. (2010) *Dominant Air Power in the Information Age* (London: International Institute for Strategic Studies), February 15th, available online at *www.iiss.org/recent-key-addresses/air-chief-marshal-sir-stephen-dalton/*.

Davis G. (2007) History of the NOAA Satellite Program, *Journal of Applied Remote Sensing*, **1** (012504), January 27th, available online at *www.osd.noaa.gov/ download/JRS012504-GD.pdf*.

Day D. A., Logsdon J. M., and Ladell B. (eds) (1998) *Eye in the Sky* (Washington, DC: Smithsonian Institution Press).

de Selding P. (2004) German Military Prepares for 2005 SAR-Lupe Deployment, *Space News*, May 24th, 6.

de Selding P. B. (2007a) France Sets its Sights on an Operational Elint System, *Space News*, June 18th.

de Selding P. B. (2007b) Germany's TerraSAR-X Launched, *Space News*, June 18th, 6.

de Selding P. B. (2008) Pixel Factory Provides Increasingly Popular Cheap and Easy Imaging, *Space News*, November 24th, 17.

de Selding P. B. (2009a) Astrium Inks Two-Satellite Deal with Kazakhstan, *Space News*, October 12th.

de Selding P. B. (2009b) GeoEye Earnings Sharply Higher on GeoEye-1 Sales, *Space News*, November 16th, 10.

de Selding P. B. (2009c) Spain Presses Ahead with Dual-Use Imaging Satellites, *Space News*, November 2nd, 11.

de Selding P. B. (2010a) Controversy Deepens Over European Weather Satellite Contract, *Space News*, March 15th, 14–18.

de Selding P. B. (2010b) EMS Enters Personnel Security Market with New Tracking Device, *Space News*, April 12th, 28.

de Selding P. B. (2010c) Eumetsat Fails to Win Approval for Weather Satellite Program, *Space News*, March 29th.

de Selding P. B. (2010d) French Helios 2B Spy Sat Sends Back First Test Images, *Space News*, January 4th, 16.

de Selding P. B. (2010e) GeoEye Takes on Equity Investor for NGA Contract, *Space News*, March 15th, 6–10.

de Selding P. B. (2010f) Helios 2B Launch Lessens Pressure on Finding MUSIS Solution, *Space News*, January 11th, 10.

Dorale J. M. *et al.* (2010) Sea-Level Highstand 81,000 Years Ago in Mallorca, *Science*, February 12th, **327**(5967), 860–863.

Doyle J. M. (ed.) (2008) Radar Love, *Aviation Week & Space Technology*, September 29th, 23.

Farr, T. G. *et al.* (2007) The Shuttle Radar Topography Mission, *Rev. Geophys.*, **45**, RG2004, doi: 10.1029/2005RG000183.

Fiennes R. (2006) *Above the World: Stunning Satellite Images from Above Earth* (London: Cassell Illustrated), Foreword.

Fischer D. (1997) *History of the International Atomic Energy Agency, the First Forty Years* (Vienna, Austria: IAEA).

Fischlin A. *et al.* (2007) Ecosystems, their Properties, Goods, and Services, in Parry M. L. *et al.* (eds), *Climate Change 2007: Impacts, Adaptation and Vulnerability. Contribution of Working Group II to the Fourth Assessment Report of the Intergovernmental Panel on Climate Change* (Cambridge, UK: Cambridge University Press), 211–272.

Foley J. (2010) Boundaries for a Healthy Planet, *Scientific America*, April, **302**(4), 38–41.

Friis-Christensen E. and Lassen K. (1991) Length of the Solar Cycle: An Indicator of Solar Activity Closely Associated with Climate, *Nature*, **254**, 698–700.

Fulghum D. A. and Butler A. (2007) Reassessing Space: US Analysts Sort through the Fallout from China's Satellite Shoot-Down, *Aviation Week & Space Technology*, April 30th, 27–29.

Fyler T. (ed.) (2010) GLONASS to Go Commercial, Says Putin, *Navigation News*, March/April, 4.

Gates R. M. (2010) *Nuclear Posture Review Report*, US Department of Defense, April, Preface.

Gee A. (2009) And Then They Came for the Jews, *The (London) Sunday Times Magazine*, November 1st, 24–31.

Goldberg J. (2009) How Iran Could Save the Middle East, *The Atlantic*, July/August.

Gorman S. (2008) Satellite-Surveillance Program to Begin Despite Privacy Concerns, *The Wall Street Journal*, October 1st.

Gorman S. (2009) White House to Abandon Spy-Satellite Program, *The Wall Street Journal*, June 23rd.

Grimston J. and Haslam C. (2010) This Is Just the Beginning, Warn Scientists, *London Sunday Times*, April 18th, 10.

Grishin V. (2009) [Federal Space Agency] FSA Report to the CCSDS Management Council, October (should in due course be available online at *http://cwe.ccsds.org/cmc/docs/Forms/AllItems.aspx?RootFolder=%2Fcmc%2Fdocs&View=%7BD5DD30F7%2D53FC%2D45B9%2D8B93%2D709B280A475B%7D*).

Grzebellus M. (2010) Being under Watch, *Coordinates*, April, 12–16.

Hanley C. (2000) Regulating Commercial Remote Sensing Over Israel: A Black Hole in the Open Skies Doctrine? *Administrative Law Review*, Winter, **52**, 423–442.

Harding T. (2007) Terrorists 'Use Google Maps to Hit UK Troops', available online at *www.telegraph.co.uk/news/worldnews/1539401/Terrorists-use-Google-maps-to-hit-UK-troops.html*, January 13th.

Harris R. (2009) *Remote Sensing Policy*, in Warner T., Ellis M., and Foody G. (eds), *The SAGE Handbook of Remote Sensing* (London: SAGE Publications), **2**, 18–29.

Hastings M. (2008) *Nemesis: The Battle for Japan, 1941–45* (London: Harper Perennial).

Hayes J. (2008) Bridging Research and Operations for Future Environmental Services, 88th AMS Annual Meeting, January 22nd, available online at *www.weather.gov/com/files/GEOSS_Oct_NN.pdf*.

Hayes J. (2009) NOAA National Weather Service Serving the Nation's Environmental Forecasting Needs, available online at *www.nws.noaa.gov/com/files/lubchenco_brief_040309.pdf*, April 3rd.

Hayes-Ryan K. (2009) Keynote Speech at MILSPACE 2009, Paris, France, April 27–28th, available online at *www.nro.gov/Speeches/MILSPACE_2009.pdf*.

Hendrickx B. (2005) Snooping on Radars: A History of Soviet/Russian Global Signals Intelligence Satellites, *Journal of the British Interplanetary Society*, **58**(Suppl. 2), 97–133.

Heyman J. (2009) EW, ELINT & Surveillance Satellites, *Milsat Magazine*, January, 21–32.

Hsu S. S. (2009) DHS to Cut Police Access to Spy-Satellite Data, *Washington Post*, June 24th.

IAEA Secretariat (2006) *Strengthening the Effectiveness and Improving the Efficiency of the Safeguards System Including Implementation of Additional Protocols* (Vienna, Austria: IAEA), GC(50)2, August 7th.

IAEA Secretariat (2007) *IAEA Safeguards: Staying Ahead of the Game* (Vienna, Austria: IAEA), July.

IAEA Secretariat (2009) *Annual Report 2008* (Vienna, Austria: IAEA), GC(53)7, September.

Iannotta B. (2008) Lockheed Martin Lands $1 Billion Weather Satellite Contract, *Space News*, December 8th, 14.

Iannotta B. (2009) Pentagon Eyes Satellite Equipped with Joystick-Controlled Video Cameras, *Space News*, November 23rd, 12.

Indian Government (2009) Dossier Presented to Pakistan Government on Mumbai Terrorist Attacks (Nov. 26–29, 2008), online edition of *The Hindu*, *www.hindu.com/nic/dossier.htm*, January 5th.

Inmarsat (2010) *Via Inmarsat*, January–March.

ITU (2010) ITU Sees 5 Billion Mobile Subscriptions Globally in 2010: Strong Global Mobile Cellular Growth Predicted across All Regions and All Major Markets, International Telecommunications Union Press Release, February 15th, available online at *www.itu.int/net/pressoffice/press_releases/2010/06.aspx*.

Jansen E. J. *et al.* (2007) Palaeoclimate, in Solomon S. *et al.* (eds), *Climate Change 2007: The Physical Science Basis. Contribution of Working Group I to the Fourth Assessment Report of the Intergovernmental Panel on Climate Change* (Cambridge, UK and New York: Cambridge University Press).

Jaroslovsky R. (2010) In GPS, Google Is Still a Little Lost, *Bloomberg Business Week*, January 18th, 69.

Jayaraman K. S. (2008) ISRO Helps Indian Military Establish Space Cell, *Space News*, June 23rd, 8.

Johansen A. (1995) Riesige Lauscher am Himmel, *Die Zeit*, July 28th.

Kallender-Umezu P. (2009) North Korean Missile Test Puts Focus on Japanese Space Policy, *Space News*, April 13th, 7.

Kehler R. (2003) *Space in US Military Strategy*, in *The Impact of Space upon Military Operations* (London: Royal United Services Institute), June 12th–13th.

Keith A. (2010a) Diversifying Capabilities for Image Intelligence, *Space News*, July 19th, 16–1.

Keith A. (2010b) Low-Cost Data: Threat to Commercial Operators? *Space News*, January 11th, 19–21.

Kennett D. J. *et al.* (2009), Nanodiamonds in the Younger Dryas Boundary Sediment Layer, *Science*, January 2nd, **323**(5910), 94.

Kim T.-H. (2010) Space Program Faces Crucial Test this Year, available online at *www.ecognition.cc/bbs/dv_rss.asp?s = xhtml&boardid = 13&id = 478&page = 1*, February 20th.

Kleiner K. (2009) Mind What You Say on your Mobile, *New Scientist*, December 18th.

Langewiesche W. (2005) The Wrath of Khan: How A. Q. Khan Made Pakistan a Nuclear Power and Showed that the Spread of Atomic Weapons Can't Be Stopped, *The Atlantic*, **296**(4), 62–85.

Lardier C. (2006) Le budget spatial militaire américain 2007, *Air & Cosmos*, February 17th, **2018**, 36.

Lardier C. (2007) Succès pour Cosmo/Skymed-1, *Air & Cosmos*, June 15th, **2082**, 152.

Lardier C. (2008a) Espace militaire français: espoirs et incertitudes, *Air & Cosmos*, November 7th, **2146**, 36–37.

Lardier C. (2008b) Le satellite Pléiades-1 sort de ses difficultés: Fin juillet, Thales Alenia Space va livrer l'instrument optique à EADS Astrium Toulouse pour son integration sur la platform. L'ensemble sera livré à Kourou dans quinze mois, *Air & Cosmos*, July 11th, **2133**, 60–61.

Lardier C. (2008c) Neuf tirs dont deux échecs en juillet et août, *Air & Cosmos*, August 29th, **2136**, 32–33.

Lardier C. (2008d) Satellites de Chine, d'Inde et d'Israel, *Air & Cosmos*, May 2nd, **2123**, 64.

Lardier C. (2008e) Un Cosmos d'alerte avancée russe, *Air & Cosmos*, July 4th, **2132**, 35.

Lardier C. (2009a) Bientôt, un commandement pour l'espace militaire, *Air & Cosmos*, December 11th, **2198**, 38–39.

Lardier C. (2009b) Cosmos-2455, *Air & Cosmos*, November 27th, **2196**, 54.

Lardier C. (2009c) Nouveautés spatiales à Joukovski: Le future lanceur Rus-M et le successeur du Soyouz-TMA étaient presents par Energya et TsSKB-Progress, *Air & Cosmos*, August 28th, **2183**, 18–19.

Lardier C. (2009d) Succès pour le satellite-espion américain, *Air & Cosmos*, January 23rd, **2156**, 38–39.

Lardier C. (2009e) Un mystérieux satellite chinois, *Air & Cosmos*, November 20th, **2195**, 40.

Lardier C. (2009f) Une Delta-4H pour NRO-26, *Air & Cosmos*, January 9th, **2154**, 35.

Lardier C. (2010a) La Chine lance un nouveau Yaogan: Il s'agit probablement d'un satellite d'imagerie radar militaire. Par ailleurs, la Chine a reporté le lancement du module orbital Tiangong-1 à 2011, *Air & Cosmos*, March 12th, **2209**, 38–39.

Lardier C. (2010b) La Russie lève le voile sur Orletz, *Air & Cosmos*, April 30th, **2216**, 70.

Lardier C. (2010c) La Russie maintient son avance en 2009, *Air & Cosmos*, January 8th, **2200**, 26–32.

Lavers C. (2010) Remote Sensing Monitors Displaced Persons, *GEOconnexion International*, April, **9**(4), 26–28.

Leake J. (2010) Arctic Ice Recovers from the Great Melt, *The (London) Sunday Times*, April 4th, 8.

Léon J.-C. (2010) Drones à l'attaque, *Air & Cosmos*, February 26th, **2207**, 9.

Leslie R., Riggs P., Bragini V., Truong Q., Neville R., and Staenz K. (2002) Satellite Imagery for Safeguards Purposes: Utility of Panchromatic and Multispectral Imagery for Verification of Remote Uranium Mines, *Annual Meeting of the Institute of Nuclear Materials Management*, June, 23–27.

Levy C. J. (2010) Ukraine Woos Russia with Lease Deal, *New York Times*, April 22nd, New York edition, A8.

Lindgren D. T. (2000) *Trust but Verify* (Annapolis, MD: Naval Institute Press).

Little J. B. (2009) Saving the Ogallala Aquifer, *Scientific American Earth 3.0*, March, **19**(1), 32–39.

Longhorn R. (ed.) (2010a) DigitalGlobe Today and Tomorrow, *Geo:Connexion International*, May, **9**(5), 5.

Longhorn R. (ed.) (2010b) Solving Precision Farming Challenges, *Geo:Connexion International*, May, **9**(5), 20–23.

Longhorn R. (ed.) (2010c) WorldView-2 Elevation Data Accuracy Verified, *Geo:Connexion International*, May, **9**(5), 10.

Marcelo E., Kuhn J., and Bush R. I. (2010) One Solar Cycle of Solar Astrometry with MDI/SOHO, in *Solar and Stellar Variability: Impact on Earth and Planets*, Proceedings of the International Astronomical Union, IAU Symposium, February, **264**, 21–32.

Mason P. J. and Bojinski S. (eds) (2006) *Systematic Observation Requirements for Satellite-Based Products for Climate: Supplemental Details to the GCOS Implementation Plan* (Geneva, Switzerland: World Meteorological Organisation), GCOS-107 (WMO/TD No. 1338), September, available online at *www.wmo.int/ pages/prog/gcos/Publications/gcos-107.pdf*.

Massom R. and Lubin D. (2006) *Polar Remote Sensing, Volume II: Ice Sheets* (Chichester, UK: Springer-Praxis).

Matthews W. (2009) TacSat-3 Earth Observing Satellite Puts Hyperspectral Imagery Analysis in Space, *Space News*, June 22nd, 16.

McDonald R. A. (ed.) (2002) *Beyond Expectations—Building an American National Reconnaissance Capability: Recollections of the Pioneers and Founders of National Reconnaissance* (Washington, DC: American Society for Photogrammetry & Remote Sensing).

McDowell J. (2010) *Jonathan's Space Report*, April 6th, **625**, available online at *http://host.planet4589.org/space/jsr/back/news.625*.

McEntire M. (2010) Ensnared by Error on Growing US Watch List, *New York Times*, April 7th, A1.

Mecham M. (2010) First Mirror Shines: Polished to a 20-Nanometer Tolerance, JWST's Segment Passes Big Cryo Test, *Aviation Week & Space Technology*, March 8th, 36.

Meehl G. A. *et al.* (2007), Global Climate Projections, in Solomon S. *et al.* (eds), *Climate Change 2007: The Physical Science Basis. Contribution of Working Group I to the Fourth Assessment Report of the Intergovernmental Panel on Climate Change* (Cambridge, UK and New York: Cambridge University Press).

Mesrine M. (ed.) (2008) Final Report: Applications for Aperture Synthesis Techniques for Imaging in Earth Observation and Science, *Thales Alenia Space*, ASS-Appli-TN-08, October 13th.

Mikoyan S. A. (2001) Eroding the Soviet "Culture of Secrecy": Western Winds Behind Kremlin Walls, *Studies in Intelligence*, available online at *www.cia.gov/ library/center-for-the-study-of-intelligence/csi-publications/csi-studies/studies/fall_ winter_2001/article05.html*.

Morring F. Jr (ed.) (2007a) Russian Milsats, *Aviation Week & Space Technology*, September 24th, 23.

Morring F. Jr (ed.) (2007b) Watching North Korea, *Aviation Week & Space Technology*, March 5th, 15.

Morring F. Jr (ed.) (2009) Overhead Reconnaissance, *Aviation Week & Space Technology*, December 7th, 19.

Morring F. Jr (ed.) (2010a) Climate Cooperation, *Aviation Week & Space Technology*, April 19th, 18.

Morring F. Jr (ed.) (2010b) Israel Launch Aims to Boost Intel on Iran, *Aviation Week & Space Technology*, February 15th, 16.

Moskvitch K. (2010) Glonass: Has Russia's Sat-Nav System Come of Age? BBC News Online, *http://news.bbc.co.uk/1/hi/sci/tech/8595704.stm*, April 2nd.

Musquère A. (2009) L'espace au service du vignoble, *Air & Cosmos*, October 2nd, **2188**, 44.

Naff C. F. (2010) Geologists Drill into Antarctica and Find Troubling Signs for Ice Sheets' Future, *Scientific American Online*, *www.scientificamerican.com/article.cfm?id=antarctica-andrill-ice-sheets*, April 19th.

NASA (2003) NASA Study Finds Increasing Solar Trend That Can Change Climate, NASA website, *www.nasa.gov/centers/goddard/news/topstory/2003/0313irradiance.html*.

NASIC (National Air and Space Intelligence Center) (2009) *Ballistic and Cruise Missile Threat* (Wright-Patterson AFB, Ohio: NASIC), NASIC-1031-0985-09, April, available online at *www.fas.org/irp/threat/missile/naic/NASIC2009.pdf*.

National Research Council (2010) *Defending Planet Earth: Near-Earth Object Surveys and Hazard Mitigation Strategies: Final Report* (Washington, DC: The National Academies Press), pre-publication copy, January.

Nativi A. (2007) Cosmo Calling: With X-Band Radar Safely in Orbit, Italy Starts Work on Second-Generation Design, *Aviation Week & Space Technology*, June 18th, 72.

Nativi A. and Taverna M. (2009) Turkish Delight: Telespazio to Supply Gokturk Surveillance Satellite, US Milsatcom Capacity, *Aviation Week & Space Technology*, January 5th, 31–33.

New Scientist (2009a) Ice Loss in Greenland behind a Sixth of Global Sea-Level Rise, November 21st, 17.

New Scientist (2009b) US Plays "I Spy a Broken Sat", January 24th, 21.

Niemeyer I. (2009) *Perspectives of Satellite Imagery Analysis for Verifying the Nuclear Non-Proliferation Treaty*, in Jasani B., Niemeyer I., Nussbaum S., Richter B., and Stein D. (eds), *International Safeguards and Satellite Imagery: Key Features of the Nuclear Fuel Cycle and Computer-Based Analysis* (Berlin-Heidelberg: Springer-Verlag).

Noren J. (2007) CIA's Analysis of the Soviet Economy, in Haines G. K. and Leggett R. E. (eds), *Watching the Bear: Essays on CIA's Analysis of the Soviet Union* (Washington, DC: CIA/CSI), available online at *www.cia.gov/library/center-for-the-study-of-intelligence/csi-publications/books-and-monographs/watching-the-bear-essays-on-cias-analysis-of-the-soviet-union/article02.html*.

Norris P. (2007) *Spies in the Sky* (Chichester, UK: Springer-Praxis).

Norris R. S. and Kristensen H. M. (2008) Chinese Nuclear Forces, 2008, *Bulletin of the Atomic Scientists*, July/August, 42–45.

Norris R. S. and Kristensen H. M. (2010) Russian Nuclear Forces, 2010, *Bulletin of the Atomic Scientists*, January/February, 74–81.

Northern Sky Research (NSR) (2009) *Global Satellite-Based Earth Observation: An Assessment of Growth Potential for 2008–2018*, available online at *www.nsr.com/index.php?option=com_virtuemart&Itemid=134*, November.

O'Brien K. (2009) Cellphone Encryption Code Is Divulged, *New York Times*, December 28th.

Opall-Rome B. (2008) 1st Israeli Radar Imaging Satellite to Upgrade Strategic Intel Capabilities, *Space News*, January 28th, 5.

Opall-Rome B. (2009) Ofeq-8 Nearing Launch, Ofeq-9 Stalled, *Space News*, November 16th, A3.

Opall-Rome B. (2010) Israeli Missile Experts: Simorgh Sets Iran on Path to ICBM, *Space News*, February 15th, 14.

Pachauri R. K. and Reisinger A. (eds) (2007) *Climate Change 2007: Synthesis Report. Contribution of Working Groups I, II and III to the Fourth Assessment Report of the Intergovernmental Panel on Climate Change* (Geneva, Switzerland: IPCC).

Parkinson B. (2010) His Coordinates: Dr Bradford W Parkinson, Chief Architect of Global Positioning System, *Coordinates*, January, VI(1), 7–11.

Parmalee P. J. (ed.) (2010) Some Shut-Eye for RapidEye, *Aviation Week & Space Technology*, February 1st, 16.

Petras C. M. (2005) "Eyes" on Freedom: A View of the Law Governing Military Use of Satellite Reconnaissance in US Homeland Defense, *Journal of Space Law*, **31**(1), 81–115.

Pincher C. (1984) *Too Secret Too Long: The Great Betrayal of Britain's Crucial Secrets and the Cover-Up* (London: Sidgwick & Jackson).

Pirard T. (2007) Taïwan: cap sur l'autonomie spatial, *Air & Cosmos*, November 9th, **2099**, 32.

Pirard T. (2010) Dix Beidou à lancer entre 2010 et 2012, *Air & Cosmos*, March 19th, **2210**, 38.

Posen B. R. (2000) US Security Policy in a Nuclear-Armed World, or What If Iraq Had Had Nuclear Weapons, in Utgoff V. A. (ed.), *The Coming Crisis: Nuclear Proliferation, US Interests, and World Order* (Cambridge, MA: MIT Press).

Prober R. (2003) Shutter Control: Confronting Tomorrow's Technology with Yesterday's Regulations, *Journal of Law and Politics*, Spring, **19**, 203–252.

Purdy R. (2006) Satellites: A New Era for Environmental Compliance? *Journal for European Environmental & Planning Law*, **5**, 406–413.

Rapp D. (2008) *Assessing Climate Change: Temperatures, Solar Radiation, and Heat Balance* (Chichester, UK: Springer-Praxis).

Rashid A. (2009) *Descent into Chaos: The World's Most Unstable Region and the Threat to Global Security* (London: Penguin Books).

Reed T. C. and Stillman D. B. (2009) *The Nuclear Express: A Political History of the Bomb and its Proliferation* (Minneapolis, MN: Zenith Press).

Rhodes R. (1986) *The Making of the Atomic Bomb* (New York: Simon & Schuster Paperbacks).

RIA Novosti (2010) Glonass satellite system should go commercial—Putin, *RIA Novosti*, February 15th, available online at *http://en.rian.ru/russia/20100215/157889275.html*.

Richelson J. T. (1999) *America's Space Sentinels: DSP Satellites and National Security* (Lawrence, KS: University Press of Kansas).

Richelson J. T. (2002) *The Wizards of Langley: Inside the CIA's Directorate of Science & Technology* (Boulder, CO: Westview Press).

Richelson J. T. (2008) *The US Intelligence Community*, 5th edn (Boulder, CO: Westview Press).

Roberts D. (2010) Drought in China Hits the Energy Sector, *Bloomberg Businessweek*, April 26th, 20–23.

Robock A. and Toon O. B. (2010) Local Nuclear War, Global Suffering, *Scientific American*, January, **302**(1), 60–67.

Rodell M., Velicogna I., and Famiglietti, J. S. (2009) Satellite-Based Estimates of Groundwater Depletion in India, *Nature*, August, **460**, 999–1003.

Rogers P. (2008) Facing the Freshwater Crisis, *Scientific American*, August, **299**(2), 28–35.

Santer M. J. and Seffen K. A. (2009) Optical Space Telescope Structures: The State of the Art and Future Directions, *The Aeronautical Journal*, October, **113**(1148), 633–645.

Sapp B. (2010) Statement for the Record before the House Armed Services Committee Subcommittee on Strategic Forces, April 21st, available online at *www.nro.gov/speeches/PDDNRO_SFR_HASC_21_April_2010.pdf*.

Scharroo R. and Visser P. (1998) Precise Orbit Determination and Gravity Field Improvement for the ERS Satellites, *Journal of Geophysical Research*, **103**(C4), 8113–8127.

Schmid G. *et al.* (2001) *Report on the Existence of a Global System for the Interception of Private and Commercial Communications (ECHELON Interception System)*, European Parliament Report A5-0264/2001, July 11th.

Schulte P. *et al.* (2010) The Chicxulub Asteroid Impact and Mass Extinction at the Cretaceous–Paleogene Boundary, *Science*, March 5th, **327**(5970), 1214–1218.

Schwirtz M. (2008) Russia and Ukraine Lock Horns over Naval Base, *New York Times*, May 24th.

Scutro A. (2007) Photo of Sensitive Sub Propeller Hits the Web, *Navy News*, August 10th.

Secretary of State for Defence (2010) *Adaptability and Partnership: Issues for the Strategic Defence Review* (London: HMSO), Cm 7794, February.

Shalal-Esa A. (2009) Pentagon Sees Expanding Use of Satellite Imagery, *Reuters*, October 1st, available online at *www.reuters.com/article/idUSN0153045820081002*.

Sheridan M. (2009) Vanishing Glaciers Jolt Smokestack China: Devastation on the Tibetan Plateau Has Awakened Beijing to the Threat of Climate Change, *The [London] Sunday Times*, November 8th, 35.

Shiga D. (2010) IPCC in the Firing Line over Glacier Claim, *New Scientist*, January 16th, **2743**(205), 11.

Shindell D., Rind D., Balachandran N., Lean J., and Lonergan P. (1999) Solar Cycle Variability, Ozone, and Climate, *Science*, **284**, 305–308.

Siddiqi A. (1999) Staring at the Sea: The Soviet RORSAT and EORSAT Programmes, *Journal of the British Interplanetary Society*, **52**, 397–416.

Silva M. (2009) After Cheney, Vice Presidential House Is Clearer on Google Earth, *Los Angeles Times*, January 27th.

Singh B. (2008) Israel Emerges as India's Largest R&D Partner, *Battlespace*, October, **11**(5), 6.

Smith M. (2006) Taliban Leader "Killed" after RAF Tracks Phone, *Sunday Times*, December 24th, available online at *www.timesonline.co.uk/tol/news/uk/article 1264344.ece*.

Smith T. (2004) Policing EU Farm Subsidies, the Italian Way, BBC News, September 16th, available online at *http://news.bbc.co.uk/go/pr/fr/-/1/hi/business/ 3658010.stm*.

Solomon S. *et al.* (2007) Technical Summary, in Solomon S. *et al.* (eds), *Climate Change 2007: The Physical Science Basis. Contribution of Working Group I to the Fourth Assessment Report of the Intergovernmental Panel on Climate Change* (Cambridge, UK and New York: Cambridge University Press).

Space News (2007) Japan Lofts Two Imaging Satellites for Constellation, March 5th, 4.

Space News (2010a) Lower-Cost Satellite Imagery Benefits Cash-Strapped European Data Center, April 26th, 3.

Space News (2010b) MUSIS Ground System Deal Teters on Edge of Collapse, April 26th, 6.

Space News (2010c) NGA Extends Deal with Spot Image, March 8th, 3.

Stern D. (2010) Parliamentary Chaos as Ukraine Ratifies Fleet Deal, BBC News Online, *http://news.bbc.co.uk/1/hi/8645847.stm*.

Stone R. (2008) Target Earth, *National Geographic*, August, **214**(2), 134–149.

Stryker T. S. (2009) Land Surface Imaging (LSI) Constellation, 23rd CEOS Plenary (Phuket, Thailand), December 3rd–5th, available online at *www.symbioscomms. com/ceos2009/files/CEOS23_Item16_3_23rd%20CEOS%20plenary%20 Presentation%20LSI%20constellation.ppt*.

Stuart T. (2009) *Waste: Uncovering the Global Food Scandal* (London: Penguin Books).

Sturm M. (2010) Arctic Plants Feel the Heat, *Scientific American*, May, **302**(5), 48–55.

Sullivan E. (2009) DHS to Kill Domestic Satellite Spying, *Associated Press*, June 23rd.

Tapley B. D. *et al.* (1994) Precision Orbit Determination for TOPEX/POSEIDON, *Journal of Geophysical Research*, **99**(C12)(24), 383–404.

Taverna M. A. (2008) Private Eye: German Commercial Remote Sensing Constellation Aims to Shake Up Geo-Information Business, *Aviation Week & Space Technology*, September 8th, 40–43.

Taverna M. A. (2010) Force Fusion: France's Helios IIB Satellite Reinforces European Imagery-Intelligence Capability, *Aviation Week & Space Technology*, January 4th, 37–38.

Taverna M. A. and Wall R. (2009) Altogether Now: Despite Cooperation Snags, France Presses Ahead with Ambitious Milspace Agenda, *Aviation Week & Space Technology*, December 7th, 44–45.

Thomas A., Meikle C., and Shilston D. (2010) Remote Monitoring of a Mud Volcano, *GEOconnexion International*, April, **9**(4), 36–39.

Umansky E. (2002) The World: Image Problems; A Place to Find Out for Yourself About the War, *The New York Times*, September 22nd, **4**, 4.

United Nations (2000) *United Nations Treaties and Principles on Outer Space: Text and Status of Treaties and Principles Governing the Activities of States in the Exploration and Use of Outer Space Adopted by the United Nations General Assembly* (New York: United Nations), A/AC.105/572/Rev.3, Sales No. E00.I.24.

United Nations (2009) Copenhagen Accord FCCC/CP/2009/L.7 (New York: United Nations), December 18th, available online at *http://unfccc.int/resource/docs/2009/cop15/eng/l07.pdf*.

Urban M. (1996) *UK Eyes Alpha: The Inside Story of British Intelligence* (London: Faber & Faber).

Valpolini P. (2010) ISR in Afghanistan: SR Easier than I, *Armada International*, April/May, (2), 46–50.

Vijayan J. (2010) Researchers Use PC to Crack Encryption for Next Gen GSM Networks, Computerworld, available online at *www.computerworld.com*, January 14th.

Weidensaul S. (1999) *Living on the Wind: Across the Hemisphere with Migratory Birds* (New York: North Point Press).

Weinberger S. (2008) Can You Spot the Chinese Nuclear Sub? *DISCOVER Magazine*, August, 21–26.

Williams J. (2001) *GIS Processing of Geocoded Satellite Data* (Chichester, UK: Springer-Praxis).

Willis P. (ed.) (2010) *The Disclosure of Climate Data from the Climatic Research Unit at the University of East Anglia* (London: The Stationery Office), HC387-1, March 24th.

WMO (2010) *Current Geostationary (GEO) Satellites Contributing to the GOS*, maintained by the WMO Space Programme on behalf of CGMS, available online at *www.wmo.int/pages/prog/sat/GOSgeo.html*.

Wolmar C. (2009) *Blood, Iron & Gold* (London: Atlantic Books).

Wright D., Grego L., and Gronlund L. (2005) *The Physics of Space Security: A Reference Manual* (Cambridge, MA: American Academy of Arts & Sciences).

Zandbergen R. *et al.* (1997) High-Precision Orbit Determination and Altimeter Calibration for ERS, in Guyenne T.-D. (ed.), *Proceedings of the 12th International Symposium on Space Flight Dynamics* (Paris, France: ESA), ESA SP-403, 477–484.

Index

(**boldface** indicates main entry page range)